计算机科学与技术丛书

Vue+Spring Boot前后端
分离开发实战

贾志杰 ◎ 编著
Jia Zhijie

VUE+SPRING BOOT SPLIT
DEVELOPMENT PRACTICE

清华大学出版社
北京

内容简介

本书以实战项目为主线，以理论基础为核心，引导读者渐进式学习 Vue＋Spring Boot。Vue 可以驱动采用单文件组件和 Vue 生态系统支持的库来开发复杂的单页面应用；Spring Boot 框架是目前微服务框架的最佳选择之一。Vue＋Spring Boot 的完美结合，能够让我们在开发前后端分离项目时得心应手，从而快速开发大型 SPA 应用。

本书共 18 章，分为 Vue 篇和 Spring Boot 篇。Vue 篇（第 1～9 章）详细讲述 Vue 框架的技术知识点，以及纯前端打造的项目，从 Vue 框架基本语法的使用逐步深入到 Vue 实战项目的运用；Spring Boot 篇（第 10～18 章）从零起步，系统、深入地剖析 Spring Boot 的核心知识点及 Spring Boot 整合的众多流行技术。本书示例代码丰富，实际性和系统性较强，并配有视频讲解，助力读者透彻理解书中的重点、难点。

本书不仅适合初学者入门，精心设计的案例对于工作多年的开发者也有参考价值，并可作为高等院校和培训机构相关专业的教学参考书。

本书封面贴有清华大学出版社防伪标签，无标签者不得销售。
版权所有，侵权必究。举报: 010-62782989, beiqinquan@tup.tsinghua.edu.cn。

图书在版编目(CIP)数据

Vue＋Spring Boot 前后端分离开发实战/贾志杰编著. —北京: 清华大学出版社, 2021.1(2024.1重印)
(计算机科学与技术丛书)
ISBN 978-7-302-57020-2

Ⅰ. ①V… Ⅱ. ①贾… Ⅲ. ①JAVA 语言－程序设计 Ⅳ. ①TP312.8

中国版本图书馆 CIP 数据核字(2020)第 238034 号

责任编辑: 赵佳霓
封面设计: 吴　刚
责任校对: 焦丽丽
责任印制: 宋　林

出版发行: 清华大学出版社
　网　　址: https://www.tup.com.cn, https://www.wqxuetang.com
　地　　址: 北京清华大学学研大厦 A 座　　　　　　邮　编: 100084
　社 总 机: 010-83470000　　　　　　　　　　　　　邮　购: 010-62786544
　投稿与读者服务: 010-62776969, c-service@tup.tsinghua.edu.cn
　质量反馈: 010-62772015, zhiliang@tup.tsinghua.edu.cn
　课件下载: https://www.tup.com.cn, 010-83470236
印 装 者: 北京嘉实印刷有限公司
经　　销: 全国新华书店
开　　本: 186mm×240mm　　　印　张: 24.5　　　字　数: 548 千字
版　　次: 2021 年 3 月第 1 版　　　　　　　　　　印　次: 2024 年 1 月第 6 次印刷
印　　数: 9001～11000
定　　价: 89.80 元

产品编号: 089121-01

前言
FOREWORD

不同终端的兴起,对开发人员的要求越来越高,纯浏览器端的响应式页面已经不能满足用户体验的要求,需要针对不同的终端开发不同的定制版本;为了提升开发效率,前后端分离的需求越来越被重视,前端主要负责页面的展现和交互逻辑;后端主要负责业务和数据对接,我们可以定制开发多个版本。

本书共分为 18 章,各章主要内容如下:

第 1 章介绍前端的历史,以及前端如何从静态页面过渡到如今的单页面应用及 MVVM 架构风格。

第 2 章主要介绍 Vue 是什么,以及 Vue 引入网页的方式、开发工具等。

第 3 章讲解 Vue 的基础语法。

第 4 章讲解 Vue 除了允许核心功能默认内置的指令(v-model 和 v-show),也允许注册自定义指令。

第 5 章深入讲述属性、事件和插槽这 3 个 Vue 基础概念、使用方法及其容易被忽略的一些重要细节。组件可以扩展 HTML 元素,封装可重用的代码。

第 6 章介绍 Vue 过渡和动画在恰当的时机添加或删除 CSS 类名。

第 7 章介绍 Vue 工程化,在脚手架工具的基础上进行二次开发,为项目的开发提供了很好的底层内容。

第 8 章主要讲解 Vue-router 客户端路由,使用它处理客户端请求路径、代码执行和数据展示等问题。Axios 是一个基于 promise 的 HTTP 库,在 Vue 中 Axios 是比较常用的网络请求方法。Vuex 状态管理能够在一个集中的空间处理应用状态。

第 9 章通过移动端百度音乐实战项目让读者提前体验前后端分离的魅力。

第 10 章讲解 Spring Boot 开发环境、IDEA 构建 Spring Boot 项目及 Spring Boot 目录结构等。

第 11 章介绍 Thymeleaf 模板引擎,以及 Spring Boot 异常的处理等。

第 12 章介绍在实际项目中的应用开发,如文件上传与下载、定时器及 E-mail 的发送。

第 13 章讲述 Spring Boot 的热部署和 Postman 工具。

第 14 章深入介绍 Spring Boot 集成关系数据库和非关系数据库,以及开发应用。

第 15 章介绍 Spring Boot 整合持久层技术,并进行应用开发。

第 16 章介绍 Spring Security 的基础知识,Spring Boot 如何集成 Spring Security,利用

Spring Security 实现数据库数据认证授权。

第 17 章介绍 Spring Boot 项目的打包和部署。

第 18 章为部门管理系统实战，通过实战项目对本书知识点进行应用和总结。

本书特色

企业通常要求程序员既要有实战技能，也要内功扎实，对于新项目可以快速上手，熟悉底层原理后还应后劲十足，因此在笔试和面试时结合底层知识、实战应用、设计思维三方面进行考查。针对这 3 个方面的需求，我编写了本书。本书有三大特点。

第一，注重实战应用。本书精心设计的案例对于工作多年的人也具有很高的参考价值；书中引入了两个项目，让读者体验"编程之美"和"编程之乐"。

第二，理论讲解丰富。深入浅出，环环相扣，让读者更容易理解。

第三，设计者思维植入。本书可以让读者从知其然进化到知其所以然。一名优秀的程序员不仅要有良好的编码能力，还要有对整个项目的设计思想和把控能力，为以后的发展铺下"高速公路"。

在实战情景中学习，学完即知如何快速应用到实际工作中。

读者定位

本书系统讲述基础知识和实战项目，适合 Vue 初学者、前端工程师、前后端分离爱好者，以及 Java 语言开发人员、Spring Boot 开发人员等阅读，希望此书能给大家带来实用价值。

源代码和教学视频下载

本书教学课件（PPT）及源代码请扫描下方二维码下载。扫描书中二维码可观看本书配套视频讲解。

教学课件

本书源代码

由于编者水平有限，书中难免存在疏漏，敬请读者批评指正。

贾志杰

2021 年 1 月

目录
CONTENTS

Vue 篇

第 1 章 大前端时代 ··· 3
- 1.1 网页设计发展历史 ·· 3
  - 1.1.1 传统网页开发 ··· 4
  - 1.1.2 新前端网页开发 ·· 6
- 1.2 MVVM 风格架构 ··· 7
  - 1.2.1 为什么会出现 MVVM ·· 7
  - 1.2.2 MVVM 架构的最佳实践 ·· 9
  - 1.2.3 MVC、MVP 和 MVVM 开发模式比较 ························ 9

第 2 章 认识 Vue.js ··· 12
- 2.1 Vue 简述 ··· 12
  - 2.1.1 什么是 Vue ··· 12
  - 2.1.2 为什么选择 Vue ··· 13
- 2.2 Vue 的三种安装方式 ·· 14
- 2.3 Vue 开发工具 ·· 15
- 2.4 第一个 Vue 程序 ·· 17

第 3 章 Vue 基础语法 ·· 19
- 3.1 模板语法 ··· 19
  - 3.1.1 插值 ·· 19
  - 3.1.2 指令 ·· 27
  - 3.1.3 过滤器 ··· 28
- 3.2 实例及选项 ·· 31
  - 3.2.1 数据选项 ·· 31
  - 3.2.2 属性选项 ·· 32

3.2.3　方法选项 ·· 33
　　3.2.4　计算属性 ·· 34
　　3.2.5　表单控件 ·· 39
　　3.2.6　生命周期 ·· 45
3.3　模板渲染 ·· 54
　　3.3.1　条件渲染 ·· 54
　　3.3.2　列表渲染 ·· 58
　　3.3.3　template 标签用法 ··· 60
3.4　事件绑定 ·· 61
　　3.4.1　基本用法 ·· 61
　　3.4.2　修饰符 ·· 64
3.5　基础 demo 案例 ··· 67
　　3.5.1　列表渲染 ·· 67
　　3.5.2　功能实现 ·· 68

第 4 章　自定义指令 ·· 71
4.1　指令的注册 ·· 71
4.2　指令的定义对象 ·· 72
4.3　指令实例属性 ·· 73
4.4　案例 ··· 75
　　4.4.1　下拉菜单 ·· 75
　　4.4.2　相对时间转换 ·· 77

第 5 章　组件 ·· 81
5.1　什么是组件 ·· 81
5.2　组件的基本使用 ·· 83
　　5.2.1　全局注册 ·· 83
　　5.2.2　局部注册 ·· 86
　　5.2.3　DOM 模板解析说明 ·· 87
5.3　组件选项 ·· 89
　　5.3.1　组件 props ·· 89
　　5.3.2　props 验证 ·· 92
　　5.3.3　单向数据流 ·· 94
5.4　组件通信 ·· 100
　　5.4.1　自定义事件 ·· 100
　　5.4.2　$ emit/ $ on ·· 102

5.5 内容分发 ·· 104
　　5.5.1 基础用法 ·· 104
　　5.5.2 编译作用域 ··· 107
　　5.5.3 默认 slot ·· 108
　　5.5.4 具名 slot ·· 109
　　5.5.5 作用域插槽 ··· 110
5.6 动态组件 ·· 112
　　5.6.1 基本用法 ·· 112
　　5.6.2 keep-alive ·· 113
　　5.6.3 activated 钩子函数 ··· 115
　　5.6.4 异步组件 ·· 117
　　5.6.5 ref 和 $refs ··· 118
5.7 综合案例 ·· 121

第 6 章 过渡与动画 ·· 125

6.1 元素/组件过渡 ·· 125
6.2 使用过渡类实现动画 ·· 127
　　6.2.1 CSS 过渡 ·· 127
　　6.2.2 CSS 动画 ·· 131
　　6.2.3 自定义过渡的类名 ·· 133
　　6.2.4 CSS 过渡钩子函数 ·· 134

第 7 章 前端工程化 ·· 138

7.1 Vue-cli ·· 138
　　7.1.1 Node.js ·· 138
　　7.1.2 NPM ··· 140
　　7.1.3 基本使用 ·· 141
7.2 项目打包与发布 ·· 147
　　7.2.1 使用静态服务器工具包发布打包 ··· 148
　　7.2.2 使用动态 Web 服务器（Tomcat）发布打包 ······························· 150
7.3 Vue-devtools ·· 151
　　7.3.1 Vue-devtools 的安装 ··· 152
　　7.3.2 Vue-devtools 使用 ·· 152

第 8 章 UI 组件库和常用插件 ··· 154

8.1 Element-ui ··· 154

8.2 Vue-router ... 158
8.2.1 基本用法 ... 159
8.2.2 跳转 ... 164
8.2.3 路由嵌套 ... 169
8.2.4 路由参数传递 ... 172
8.3 Axios ... 174
8.3.1 基本使用 ... 175
8.3.2 json-server 的安装及使用 ... 177
8.3.3 跨域处理 ... 181
8.3.4 Vue 中 Axios 的封装 ... 184
8.4 Vuex ... 186
8.4.1 初识 Vuex ... 186
8.4.2 基本用法 ... 189
8.4.3 模块组 ... 193

第 9 章 实战：百度音乐项目（▶ 160min） ... 196
9.1 音乐列表 ... 196
9.1.1 跨域配置 ... 199
9.1.2 音乐列表导航栏 ... 202
9.2 歌手信息 ... 205
9.3 歌曲播放 ... 212
9.4 轮播图 ... 216
9.5 搜索实现 ... 220

Spring Boot 篇

第 10 章 进入 Spring Boot 世界 ... 225
10.1 Spring Boot 简介 ... 225
10.2 Spring Boot 环境准备 ... 226
10.2.1 JDK 环境 ... 226
10.2.2 开发工具 IDEA ... 226
10.2.3 安装与配置 Maven ... 226
10.3 Spring Boot 的三种创建方式 ... 228
10.3.1 在线创建 ... 228
10.3.2 通过 Maven 创建 ... 228
10.3.3 使用 Spring Initializer 快速创建 ... 230

- 10.4 Spring Boot 项目结构介绍 · 233
  - 10.4.1 目录结构 · 233
  - 10.4.2 启动类 · 233
  - 10.4.3 POM 文件 · 234
  - 10.4.4 配置文件 · 235
- 10.5 Spring Boot 在 Controller 中的常用注解 · 236

第 11 章 Spring Boot 整合 Web 开发 · 238

- 11.1 Spring Boot 访问静态资源 · 238
- 11.2 整合 Thymeleaf · 240
  - 11.2.1 Thymeleaf 使用 · 240
  - 11.2.2 语法规则 · 243
- 11.3 Spring Boot 返回 JSON 数据 · 244
  - 11.3.1 常用数据类型转为 JSON 格式 · 245
  - 11.3.2 Jackson 中对 null 的处理 · 247
  - 11.3.3 封装统一返回的数据结构 · 247
- 11.4 Spring Boot 中的异常处理 · 250
  - 11.4.1 自定义异常错误页面 · 250
  - 11.4.2 使用@ExceptionHandler 注解处理局部异常 · 252
  - 11.4.3 使用 @ControllerAdvice 注解处理全局异常 · 253
  - 11.4.4 配置 SimpleMappingExceptionResolver 类处理异常 · 253
  - 11.4.5 实现 HandlerExceptionResolver 接口处理异常 · 254
  - 11.4.6 一劳永逸 · 255
- 11.5 配置嵌入式 Servlet 容器 · 256
  - 11.5.1 如何定制和修改 Servlet 容器的相关配置 · 256
  - 11.5.2 注册 Servlet 三大组件——Servlet、Filter、Listener · 257
  - 11.5.3 替换为其他嵌入式 Servlet 容器 · 259
- 11.6 在 Spring Boot 中使用拦截器 · 260

第 12 章 应用开发 · 263

- 12.1 文件上传与下载 · 263
  - 12.1.1 单文件上传 · 263
  - 12.1.2 多文件上传 · 265
  - 12.1.3 文件下载 · 266
- 12.2 定时器 · 267
  - 12.2.1 Task · 268

		12.2.2　Quartz ·················· 270
	12.3　Spring Boot 发送 E-mail ·················· 273
		12.3.1　发送邮件需要的配置 ·················· 273
		12.3.2　使用 Spring Boot 发送邮件 ·················· 274

第 13 章　Spring Boot 热部署和 Postman 工具 ·················· 282

	13.1　devtools 热部署 ·················· 282
		13.1.1　热部署原理 ·················· 282
		13.1.2　devtools 应用 ·················· 282
	13.2　Postman 工具 ·················· 285
		13.2.1　Postman 介绍 ·················· 285
		13.2.2　Postman 下载安装 ·················· 285
		13.2.3　Spring Boot 基于 Postman 的 RESTful 接口调用 ·················· 286

第 14 章　Spring Boot 整合数据库 ·················· 288

	14.1　非关系数据库和关系数据库的区别 ·················· 288
	14.2　整合 Redis 缓冲 ·················· 289
		14.2.1　Redis 简介 ·················· 289
		14.2.2　Redis 的安装 ·················· 290
		14.2.3　Redis 数据库操作 ·················· 291
		14.2.4　Spring Boot 整合 Redis ·················· 296
		14.2.5　Redis 缓冲在 Spring Boot 项目中的应用 ·················· 298
	14.3　整合 MongoDB ·················· 302
		14.3.1　MongoDB 简介 ·················· 302
		14.3.2　MongoDB 安装 ·················· 303
		14.3.3　常用命令 ·················· 305
		14.3.4　Spring Boot 整合 MongoDB ·················· 306
	14.4　整合 MySQL ·················· 310
		14.4.1　MySQL 简介 ·················· 310
		14.4.2　Spring Boot 整合 MySQL ·················· 310

第 15 章　Spring Boot 整合持久层技术 ·················· 314

	15.1　整合 JdbcTemplate ·················· 314
	15.2　整合 MyBatis ·················· 317
		15.2.1　MyBatis 简介 ·················· 317
		15.2.2　Spring Boot 整合 MyBatis ·················· 318

15.3 Spring Data JPA ······ 322
15.3.1 JPA、Spring Data、Spring Data JPA 的故事 ······ 322
15.3.2 整合 Spring Data JPA ······ 322
15.3.3 CORS 跨域配置 ······ 329
15.4 RESTful 风格 ······ 332

第 16 章 Spring Boot 安全框架 ······ 333
16.1 认识 Spring Security ······ 333
16.1.1 入门项目 ······ 334
16.1.2 角色访问控制 ······ 335
16.2 基于数据库的认证 ······ 336
16.2.1 Spring Security 基于数据库认证 ······ 336
16.2.2 角色访问控制 ······ 342
16.2.3 密码加密保存 ······ 343
16.2.4 用户角色多对多关系 ······ 346
16.2.5 角色继承 ······ 350

第 17 章 项目构建与部署 ······ 351
17.1 Jar 部署 ······ 351
17.2 War 部署 ······ 352

第 18 章 部门管理系统（▶ 170min）······ 354
18.1 技术分析 ······ 354
18.2 项目构建 ······ 354
18.2.1 前端项目搭建 ······ 354
18.2.2 后端项目搭建 ······ 355
18.2.3 数据库设计 ······ 355
18.3 查询数据 ······ 356
18.3.1 后端实现 ······ 356
18.3.2 前端实现 ······ 358
18.4 加载菜单 ······ 362
18.4.1 引入 ElementUI ······ 362
18.4.2 菜单 ······ 362
18.5 带分页数据查询 ······ 367
18.5.1 后端接口实现 ······ 367
18.5.2 前端实现 ······ 368

18.6 部门员工信息的录入 ·· 371
　　18.6.1 后端接口实现 ·· 371
　　18.6.2 前端实现 ·· 371
18.7 部门数据编辑 ·· 374
　　18.7.1 后端接口实现 ·· 374
　　18.7.2 前端实现 ·· 375
18.8 部门数据删除 ·· 377
　　18.8.1 后端接口实现 ·· 377
　　18.8.2 前端实现 ·· 377

Vue 篇

第 1 章　大前端时代
第 2 章　认识 Vue.js
第 3 章　Vue 基础语法
第 4 章　自定义指令
第 5 章　组件
第 6 章　过渡与动画
第 7 章　前端工程化
第 8 章　UI 组件库和常用插件
第 9 章　实战：百度音乐项目

第 1 章 大前端时代

大前端时代是 Web 统一的时代,利用 HTML5 不但可以开发传统的网站,实现炫酷的网页动态效果,还可以采用 BS 架构开发应用程序、手机端 Web 应用、移动端 Native 应用程序、智能设备(例如可穿戴智能手表、可穿戴智能衣服)等。

大前端时代的最大特点是一次开发便可同时适用于所有平台。再也不用为一个 App 而需为安卓和 iOS 两大平台分别开发而忧心了,大前端已经支持非常多的开发语言,例如 Java、PHP 等,连使用 JavaScript 制作后台都显得那么简单。

大前端这个词最早是因为在阿里内部有很多前端开发人员既写前端又写 Java 的 Velocity 模板而得来,不过现在大前端的范围已经越来越广泛了。谈到大前端,常常提及组件化、路由与解耦、工程化(打包工具、脚手架、包管理工具)、MVC 和 MVVM 架构,以及埋点和性能监控,这些知识会在这本书中一一给大家解释。

大前端不仅会成为移动开发与 Web 前端开发的发展趋势,也会成为未来的显示设备终端的开发技术趋势。大前端将做更多的终端开发、工程化等工作,而不仅仅只是开发 Web 页面。做前端开发或网页设计的人员如今肯定已经被 Vue.js 这个框架彻底"包围"了,因为它太火爆了。本章的目的就是探索 Vue.js 的出现及流行的原因。

1.1 网页设计发展历史

1. 网页早期

1991 年 8 月,Tim Berners Lee 发布了第一个简单的基于文本并包含几个链接的网站。1994 年,万维网联盟(W3C)成立,此联盟将 HTML5 确立为网页的标准标记语言。这一举动阻断了任何独立公司想要开发拥有专利的浏览器和相应的程序语言的野心,因为这会对网络的完整性产生不利的影响。W3C 一直致力于确立与维护网页编程语言的标准(例如 JavaScript)。直到 1995 年,使用 Web 的活跃用户还不足 5000 万,当时基本采用超文本标记语言(HTML)控制 Web 页面。HTML 的应用范围非常有限,因为它主要用于设计以文本为主的 Web 页面。垂直滚动和标志性的蓝色下画线页面链接也是在那个时代发展起来的一项功能。

2．基于表格的设计

表格布局使网页设计师制作网站时有了更多选择。在 HTML 中表格标签的本意是为了显示表格化的数据，但是设计师很快意识到可以利用表格构造他们设计的网页，这样就可以制作较以往作品更加复杂的多栏目网页。表格布局便流行起来，它融合了背景图片切片技术，常给人以看起来较实际布局简洁得多的结构。

3．Flash 设计年代

20 世纪 90 年代末网页的特点是将 Flash 整合到网页中。Flash 是革命性的，因为它预示着网站多媒体应用的新纪元。音乐、视频和动画也被有效地整合到网站中，从而增强了网站的美观和功能。不同的颜色、交互工具和装饰都因为 Flash 而成为可能。这个时代也因此发展出明亮而往往令人分心的形象。

4．CSS 的开发和使用

层叠样式表 CSS 语言的创建是为了给 HTML 网站一些样式。虽然 Flash 将多媒体功能整合到网站方面取得了里程碑式的进展，但在网站中仍然存在一个未定义的松散结构。CSS 旨在解决这个问题，它通过定义字体、颜色和各种其他元素的使用实现这一点。因此，CSS 是一个专门用于改进网站表示方面的程序。由于 CSS 带来的改进，设计师现在开始开发模板，这标志着 Web 风格多样性的开始。

5．响应设计

在 CSS 和 HTML 占据主导地位多年之后，需要以动态方式使用这两种语言的时代到来了。2010 年迎来了第一个专为手机设计的网页的发展。随着智能手机和平板计算机的普及，网页不再仅仅是为个人计算机市场而设计的。因此，必须采用新的设计技术。响应式 Web 设计是在此期间创建的，其目的是创建使用各种设备都可以轻松使用的站点。响应式设计最终导致了平面设计。平面网站的特点是极简和有效。

1.1.1 传统网页开发

在 Web 1.0 时代，由于网页是在服务端使用动态脚本语言和模板引擎渲染出来的，所以不分前后端，JSP 是一个典型写法，JSP 和 Java 代码结合起来，刚开始确实提高了开发效率，但时间长了，大家发现 JSP 存在很多问题，大部分后端程序员不懂前端，所以开发流程一般是这样的：UI 设计人员制作高精度图片，交给静态界面开发人员；静态界面开发人员按照图片使用 HTML、CSS 进行制作，使其生成的网页与图片保持一致；开发人员需要把 HTML 界面中的静态代码改成动态页面从而和后台进行数据对接。这个过程如果只执行一次还没有问题，但是在实际项目中 UI 经常会进行调整，静态界面开发人员也会由于制作的静态界面不符合要求进行调整，而每一次调整最后都会涉及其他开发人员，造成开发人员精力的大量浪费，如图 1-1 所示。

传统的 Web 开发好比是瀑布模型，它的下一步实现完全建立在上一步的基础之上，如果上一步没有完成，下一步就没办法开展，当用户向服务器发送请求时，服务器端必须完全处理好客户端的请求，并且客户端完全接收到服务器端的响应后，才能继续进行操作。如果

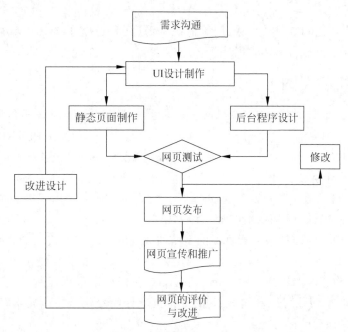

图 1-1　传统网页开发流程图

需要刷新,等待那么久必须刷新整个页面,当整个页面都加载出来时,才可以进行显示。

(1) 现时代公司的业务总会越变越复杂,这点是不可避免的,需求也是没有止境的。业务复杂度的变化会让整个系统复杂化和多元化。同时,开发团队扩张也导致参与人员很可能从几人快速扩展到几十人,提供的服务越来越多,调用关系变得复杂,前端搭建本地环境不再是一件简单的事情。不同的人提供的页面和其他人提供的页面可能会有细节上的差异,即使考虑团队协作,往往最后呈现的页面和想象中的页面也会有一些差距。

(2) 前端的样式更新操作变得复杂且会造成系统的不稳定。因为所有的页面都是基于后端而自动生成的,所以对于一些前端样式的更新和更改可能需要将整个代码逻辑重构,甚至重新上线一个崭新的系统。

(3) 随着一个项目体量的增大,其代码的维护也一定会越来越难。单一代码负责前台和后端的数据处理,会导致职责不清晰,而且由于开发人员的水平和书写习惯不同,以及各种紧急需求,糅合大量业务代码和其他历史代码,甚至意义不明的无用代码,往往会带来大量的维护成本。

(4) 特别是互联网兴起后,一套系统对应多个前端,一般除了 PC 端,还有移动的、小程序等,此时前后端不分的方式不是最优的选择,而且对于现阶段的大型应用或追求用户体验的应用而言,前后端的分离是必要的。

前后端分离的火热,得益于 Ajax 的发展,前端不再依赖后台环境生存,所有服务器数据都可以通过异步交互获取。在后台取得一个良好定义的 RESTful(Representational State Transfer,表述性状态转移)接口后,两端甚至可以在零沟通成本的情况下并行完成项目任务。

Ajax 的发展加上 CDN 开始大量用于静态资源存储，造就了 JavaScript 的火热及之后的 SPA（Single Page Application 单页面应用）时代。

1.1.2 新前端网页开发

Web 早已进入 2.0 时代，如今的网页大有向系统应用级别方向发展的趋势，再也不是以前简单展示信息的界面了。如今很多 Web App 已经做到了原生应用的功能，并且运用自身的优势逐步取代之。HTML5 也很强大，对多平台，以及多屏幕设备的良好兼容使得前端工程师在各种平台上大显身手。

单页面 SPA 应用是一种特殊的 Web 应用。它将所有的活动局限于一个 Web 页面中，仅在该 Web 页面初始化时加载相应的 HTML、JavaScript 和 CSS。一旦页面加载完成，SPA 就不会因为用户的操作而进行页面的重新加载或跳转。取而代之的是利用 JavaScript 动态地变换 HTML 的内容，从而实现 UI 与用户的交互。由于避免了页面的重新加载，SPA 可以提供较为流畅的用户体验。得益于 Ajax，我们可以实现无跳转刷新，又得益于浏览器的 histroy 机制，我们可以利用 hash 的变化推动界面变化。模拟元素客户端的单页面切换效果，如图 1-2 所示。

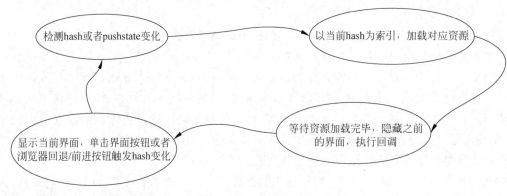

图 1-2　SPA 应用原理

一种开发模式流行起来之后，对应的框架也会随之而起。正像近几年比较流行的 Vue.js，它是目前最流行的 MVVM 框架之一，非常适合用于 SPA，本书会重点讲解 Vue.js 的基础知识和项目构建及开发。时下 SPA 单页面应用如火如荼，被人追捧是有道理的，当然它也有不足之处，正如任何东西都有两面性。

优点如下。

（1）良好的交互体验：前端进行的是局部的渲染，避免了不必要的跳转和重复的渲染。

（2）前后端职责分离（前端负责 View，后端负责 Model），架构清晰：单页 Web 应用可以和 RESTful 规约一起使用，通过 REST API 提供接口数据，并使用 Ajax 异步获取，这样有助于分离客户端和服务器的工作。

（3）减轻服务器的压力：服务器只需提供数据，而不需要管理前端的展示逻辑和页面合成，提高了性能。SPA 应用中服务器可以先将一份包含静态资源（HTML、CSS、JS 等）的静态数据（payload）发送给客户端，之后客户端只需获取渲染页面或视图数据。

（4）共用一套后端程序代码：不用修改后端程序代码就可以同时用于 Web 界面、手机、平板端等多种客户端。

缺点：

（1）SEO（搜索引擎优化）难度高：由于所有内容都在一个页面中进行动态替换，也就是利用 hash 片段实现路由，而利用 hash 片段不会作为 HTTP 请求中的一部分发送给服务器，所以在 SEO 上有着天然的弱势。而 SPA 使用 hash 片段的目的是在片段内容发送变化时浏览器不会像 URL 发送变化时那样发送请求，这样就可以只请求页面或渲染所需的数据，而不是每一个页面获取并解析整份文档。

（2）首次加载时间过长：为实现单页 Web 应用功能及显示效果，需要在加载页面时将 JS、CSS 统一加载，部分页面按需加载。

（3）页面复杂度提高，逻辑复杂程度成倍增加：由于后端只提供数据而不再管理前端的展示逻辑和页面合成，所以这些展示逻辑和页面合成都需要在前端进行编写（前进、后退等），所以会大大提高页面的复杂性和逻辑的难度。

1.2 MVVM 风格架构

MVVM 是 Model-View-ViewModel 的简写。它本质上就是 MVC 的改进版。MVVM 就是将其中的 View 的状态和行为抽象化，让我们将视图 UI 和业务逻辑分开。当然这些事 ViewModel 已经帮我们完成了，它可以取出 Model 的数据同时帮忙处理 View 中由于需要展示内容而涉及的业务逻辑。微软公司的 WPF 带来了新的技术体验，如 Silverlight、声频、视频、3D、动画……这导致了软件 UI 层更加细节化、可定制化。

同时，在技术层面，WPF 也带来了很多新特性，如 Binding、Dependency Property、Routed Events、Command、DataTemplate、ControlTemplate 等。MVVM 框架的由来便是 MVP（Model-View-Presenter）模式与 WPF 结合应用方式而发展演变过来的一种新型架构框架。它立足于原有 MVP 框架并且把 WPF 的新特性糅合进去，以应对客户日益复杂的需求变化。

1.2.1 为什么会出现 MVVM

MVC 是 Model-View-Controller 的缩写，即模型-视图-控制器，一个标准的 Web 应用程式由三部分组成。

View：用来把数据以某种方式呈现给用户。

Model：就是数据。

Controller：接收并处理来自用户的请求，并将 Model 返回给用户。

在HTML5还未流行起来的那些年，MVC作为Web应用的最佳实现方式是可以的，这是因为当时Web应用的View层相对来说比较简单，前端所需要的数据在后端基本上可以处理好，View层主要做一下展示，那时候提倡的是用Controller来处理复杂的业务逻辑，所以View层相对来说比较轻量，也就是所谓的瘦客户端思想。

相对HTML4，HTML5最大的亮点是它为移动设备提供了一些非常有用的功能，使得HTML5具备了开发App的能力，HTML5开发App最大的好处就是跨平台、快速迭代和上线，节省人力成本和提高效率，因此很多企业开始对传统的App进行改造，逐渐采用HTML5代替Native。Native使用原生系统内核，相当于直接在系统上操作，是我们传统意义上的软件，但是HTML5最大的优点是可以跨平台，开发容易，而Native需要用Android的语言和iOS的语言分别写，但HTML5只需要开发一套。到2015年，市面上大多数App或多或少都嵌入了HTML5页面。既然要用HTML5来构建App，那么View层所做的事情就不仅仅是简单的数据展示了，它不仅要管理复杂的数据状态，还要处理移动设备上各种操作行为等。因此，前端需要工程化，也需要一个类似于MVC的框架来管理这些复杂的逻辑，使开发更加高效。但这里的MVC又稍微发生了一些变化。

(1) View：UI布局，用于展示数据。

(2) Model：管理数据。

(3) Controller：响应用户操作，并将Model更新到View上。

这种MVC架构模式对于简单的应用是没有问题的，也符合软件架构的分层思想。但实际上，随着HTML5的不断发展，人们更希望使用HTML5开发的应用能和Native媲美，或者接近于原生App的体验效果，而前端应用的复杂程度已今非昔比。这时前端开发就暴露出了3个痛点问题：

(1) 开发者在代码中大量调用相同的DOM API，处理烦琐，操作冗余，使得代码难以维护。

(2) 大量的DOM操作使页面渲染性能降低，加载速度变慢，影响用户体验。

(3) 当Model频繁发生变化时，开发者需要主动更新到View；当用户的操作导致Model发生变化时，开发者同样需要将变化的数据同步到Model中，这样的工作不仅烦琐，而且很难维护复杂多变的数据状态。

MVVM由Model、View、ViewModel 3部分构成，Model层代表数据模型，也可以在Model中定义数据修改和操作的业务逻辑；View代表UI组件，它负责将数据模型转化成UI展现出来；ViewModel同步View和Model的对象。

在MVVM架构下，View和Model之间并没有直接的联系，而是通过ViewModel进行交互，Model和ViewModel之间的交互是双向的，因此View数据的变化会同步到Model中，而Model数据的变化也会立即反映到View上。

ViewModel通过双向数据绑定把View层和Model层连接起来，而View和Model之间的同步工作完全是自动的，无须人为干涉，因此开发者只需关注业务逻辑，不需要手动操作DOM，也不需要关注数据状态的同步问题，复杂的数据状态维护完全由MVVM统一管理。

1.2.2 MVVM 架构的最佳实践

MVVM 模式和 MVC 模式一样，主要目的是分离视图（View）和模型（Model），有以下几大优点。

（1）低耦合：视图（View）可以独立于 Model 变化和修改，一个 ViewModel 可以被绑定到不同的 View 上，当 View 变化的时候 Model 可以不变，同样当 Model 变化的时候 View 也可以不变。

（2）可重用性：可以把一些视图逻辑放在一个 ViewModel 里面作为可重用的控件，在具体的实例中可以引入使用，让很多 View 重用这段视图逻辑。

（3）独立开发：开发人员可以专注于业务逻辑和数据的开发（ViewModel），设计人员可以专注于页面设计，通过约束的接口规范可以进行简单的数据对接。

（4）可测试：界面向来比较难于测试，而现在测试可以针对具体的页面控件来写代码，也可以在不依赖于后端的基础上，直接通过工具或者静态数据进行测试。

1.2.3 MVC、MVP 和 MVVM 开发模式比较

1. MVC 开发模式

MVC 开发模式是 View 接收到用户的指令，传递给 Controller，然后对模型进行修改或者查找底层数据，最后把改动渲染到视图上，如图 1-3 所示。

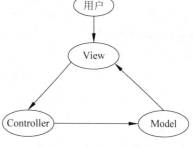

图 1-3　MVC 经典模式

MVC 的优点：

（1）耦合性低，视图层和业务层分离，这样便允许更改视图层代码而不用重新编译模型和控制器代码。

（2）重用性高。

（3）生命周期维护成本低。

（4）MVC 使开发和维护用户接口的技术含量降低。

（5）可维护性高，分离视图层和业务逻辑层也使得 Web 应用更易于维护和修改。

（6）部署快。

MVC 的缺点：

（1）不适合小型及中等规模的应用程序，花费大量时间将 MVC 应用到规模并不是很大的应用程序通常会得不偿失。

（2）视图与控制器间过于紧密连接，视图与控制器是相互分离的，但却是联系紧密的部件，视图没有控制器的存在，其应用是很有限的，反之亦然，这样就妨碍了它们的独立重用。

（3）视图对模型数据的低效率访问，依据模型操作接口的不同，视图可能需要多次调用才能获得足够的显示数据。对未变化数据的不必要的频繁访问，也将降低操作性能。

2. MVP开发模式

MVP(Model-View-Presenter)是MVC的改良模式,由IBM的子公司Taligent提出。和MVC的相同之处在于:Controller/Presenter负责业务逻辑,Model管理数据,View负责显示。另外将Controller改名为Presenter,同时改变了通信方向,如图1-4所示。

在MVP模式中,View不再负责同步逻辑,而是由Presenter负责。Presenter既负责业务逻辑也负责同步逻辑。View需要提供操作界面的接口供Presenter调用。

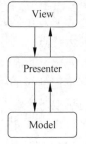

图1-4 MVP经典模式

MVP的优点:

(1) 模型与视图完全分离,可以修改视图而不影响模型。

(2) 可以更高效地使用模型,因为所有的交互都发生在一个地方——Presenter内部。

(3) 可以将一个Presenter用于多个视图,而不需要改变Presenter的逻辑。这个特性非常有用,因为视图的变化总是比模型的变化频繁。

(4) 如果把逻辑放在Presenter中,那么就可以脱离用户接口来测试这些逻辑(单元测试)。

MVP的缺点是视图和Presenter的交互过于频繁,使得它们的联系过于紧密。也就是说,一旦视图变更了,Presenter也要变更。

3. MVVM开发模式

MVVM可以看作一种特殊的MVP(Passive View)模式,或者说是对MVP模式的一种特殊改良。MVVM模式最早由微软公司提出,并且在.NET的WPF和Sliverlight中大量使用。2005年,微软公司工程师John Gossman在自己的博客上首次公布了MVVM模式。

MVVM是在原有领域Model的基础上添加一个ViewModel,这个ViewModel除了正常的属性外,还包括一些供View显示用的属性。在MVVM中,Presenter被改名为ViewModel,这样便演变成了MVVM。在支持双向绑定的平台,MVVM更受欢迎,如图1-5所示。

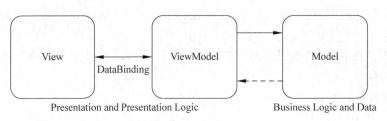

图1-5 MVVM经典模式

(1) View:封装用户界面和用户界面的任何逻辑,是一个视觉元素,定义视图及其可视布局和样式中包含的控件。

(2) ViewModel:封装视图的表示逻辑和状态,不直接引用视图,负责协调Model和View。

（3）Model：封装业务逻辑和数据，负责管理应用程序的数据并将所需的业务规则和数据验证逻辑封装，确保它的一致性和有效性。Model 类不直接引用 View 或 ViewModel，不依赖于它们的实现方式。

本书介绍的 Vue.js 是当下很流行的一个 MVVM 开发模式框架，它是以数据驱动和组件化的思想构建的。相比于 Angular.js，Vue.js 提供了更加简洁、更易于理解的 API，使得我们能够快速地掌握并使用 Vue.js。

第 2 章 认识 Vue.js

在学习 Vue 之前,读者应先学习 HTML、CSS 和 JavaScript 的基础知识。Vue 是一个专注于构建 Web 用户界面的 JavaScript 库。本章首先会对 Vue 有一个初步的介绍,然后讲解 Vue 的开发工具 WebStorm 及 Vue 的下载、使用,最后搭建 Vue 的第一个程序。

2.1 Vue 简述

Vue 在 JavaScript 前端开发库领域属于后来者,其他前端开发库有 jQuery、ExtJS、Anguals、React 等,但是 Vue 对于当前主流 JavaScript 库的地位具有很大的威胁。

2.1.1 什么是 Vue

Vue(读音/vju:/,类似于 View)是一套用于构建用户界面的渐进式框架。与其他大型框架不同的是,Vue 被设计为可以自底向上逐层应用。Vue 的核心库只关注视图层,不仅易于学习,还便于与第三方库或既有项目整合。另外,当与现代化的工具链及各种支持类库结合使用时,Vue 也完全能够为复杂的单页应用提供驱动。

Vue 的渐进式表现为:

声明式渲染→组件系统→客户端路由→大数据状态管理→构建工具。

前端框架 Vue.js 的作者尤雨溪(Evan You)是一位美籍华人,现居美国新泽西州,曾就职于 Google Creative Labs 和 Meteor Development Group。由于他在工作中大量接触开源的 Java 项目,最后自己也走上了开源之路,现在全职开发和维护 Vue.js。时至今日,Vue 已成为全世界三大前端框架之一,领先于 React 和 Angular,在国内更是首选。

Vue 重要版本发布:

(1) 2013 年,在 Google 工作的尤雨溪,受到 Angular 的启发,开发出了一款轻量框架,最初命名为 Seed。

(2) 2013 年 12 月,Seed 更名为 Vue,图标颜色采用代表勃勃生机的绿色,版本号是 0.6.0。

(3) 2014 年 1 月 24 日,Vue 正式对外发布,版本号是 0.8.0。

(4) 2014 年 2 月 25 日,0.9.0 版发布,有了自己的代号 Animatrix,此后,重要的版本都会有自己的代号。

（5）2015年6月13日，0.12.0版发布，代号 Dragon Ball，Laravel 社区（一款流行的 PHP 框架的社区）首次使用 Vue，Vue 在 JS 社区也打响了知名度。

（6）2015年10月26日，1.0.0版发布，代号 Evangelion 是 Vue 历史上的第一个里程碑。同年，Vue-router、Vuex、Vue-cli 相继发布，标志着 Vue 从一个视图层库发展为一个渐进式框架。

（7）2016年10月1日，2.0.0版发布，它是第二个重要的里程碑，它吸收了 React 的虚拟 Dom 方案，还支持服务端渲染。自从 Vue 2.0 版发布之后，Vue 就成了前端领域的热门话题。

（8）2019年2月5日，Vue 发布了 2.6.0版，这是一个承前启后的版本，在它之后，将推出 3.0.0 版。

（9）2019年12月5日，在万众期待中，尤雨溪公布了 Vue 3 源代码，目前 Vue 3 处于 Alpha 版本。

本书以 Vue 框架的稳定版本 2.9.6 进行讲解。

现在的 Vue 与运行初期相比，最大的区别就是框架涵盖的范围变大了许多。一开始 Vue 只有一个核心库，现在则包含了路由、状态管理、CLI 工具链、浏览器开发者插件、ESLint 插件等全套设施。

Vue 的定位就是为前端开发提供一个低门槛、高效率，但同时又能够伴随用户成长的框架。所谓的"伴随用户成长"，就是当一个新手用户入门的时候，Vue 尽可能地让这个过程简单直接，而当用户开始做更复杂的应用时，也就是有更复杂的需求时，用户会发现 Vue 依然能够提供良好的支持。这样 Vue 便可以在新手成长到进阶的开发者的路上一直提供所需的价值。

2.1.2　为什么选择 Vue

近两年，前端技术变革速度非常快，Vue 不论针对 Web 项目开发、网站制作，还是针对 App、小程序开发都越来越流行，其便捷性及易用性都让大家不得不考虑去学习。如果你开发的项目数据交互较多，并且前后端分离明显，那么 Vue 将会是你未来技术长足成长的不二选择。它之所以非常流行，是因为有以下几个突出的优点。

1．轻量级框架

只关注视图层，是一个构建数据的视图集合，大小只有几十 KB，Vue.js 通过简洁的 API 提供高效的数据绑定和灵活的组件系统。

2．简单易学

Vue 由华人开发，文档也为中文，不存在语言障碍，易于理解和学习。

3．双向数据绑定

双向数据绑定也就是响应式数据绑定。这里的响应式不是 @media 媒体查询中的响应式布局，而是指 Vue.js 会自动对页面中某些数据的变化做出同步的响应。也就是说，Vue.

js会自动响应数据的变化情况，并且根据用户在代码中预先写好的绑定关系，对所有绑定在一起的数据和视图内容进行修改。而这种绑定关系，就是以input标签的v-model属性来声明的，因此你在别的地方可能也会看到有人粗略地称Vue.js为声明式渲染的模板引擎。

这也就是Vue.js最大的优点，通过MVVM思想实现数据的双向绑定，让开发者不用再操作DOM对象，从而有更多的时间去思考业务逻辑。

4. 组件化

在前端应用时，我们是否也可以像编程一样把模块进行封装呢？这就引入了组件化开发的思想。

Vue.js通过组件，把一个单页应用中的各种模块拆分到一个一个单独的组件（component）中，我们只要先在父级应用中写好各种组件标签，并且在组件标签中写好要传入组件的参数（就像给函数传入参数一样，这个参数叫作组件的属性），然后再分别写好各种组件的实现，这样整个应用就算做完了。

5. 视图、数据和结构分离

使数据的更改更为简单，不需要进行逻辑代码的修改，只需操作数据就能完成相关操作。

6. 虚拟DOM

现在的网速越来越快了，很多人家里都是几十甚至上百兆位每秒的光纤，手机也是4G网络起步了，按道理一个网页才几百KB数据，而且浏览器本身还会缓存很多资源文件，那么几十兆位每秒的光纤为什么打开一个之前已经打开过的有缓存的页面还是很慢呢？这就是因为浏览器本身处理DOM也是有性能瓶颈的，尤其是在传统开发中，用JQuery或者原生的JavaScript DOM操作函数对DOM进行频繁操作的时候，浏览器要不停地渲染新的DOM树，从而导致页面看起来非常卡顿。

Virtual DOM是虚拟DOM的英文，简单来说，它可以预先通过JavaScript进行各种计算，把最终的DOM操作计算出来并优化，由于这个DOM操作属于预处理操作，所以并没有真实地操作DOM，因此叫作虚拟DOM。最后，计算完毕才真正将DOM操作提交，并将DOM操作变化反映到DOM树上。

7. 运行速度更快

相比较于React而言，同样都是操作虚拟DOM，就性能而言，Vue存在很大的优势。

2.2 Vue的三种安装方式

1. 独立版本

可以在Vue.js的官网上直接下载Vue.js文件，Vue.js官网提供了两个版本，一个是开发版本，另一个是生产版本，如图2-1所示。

注意：在开发中建议大家选用开发版本，在控制台有错误提示信息，方便调试程序。

图 2-1　Vue 下载版本

单击"开发版本"按钮，Vue.js 文件会下载到本地，直接使用<script>标签引入即可，格式代码如下：

```
<script src = "文件路径/Vue.js"></script>
```

2. 使用 CDN 方法（初学者使用）

也可以直接使用 CDN 的方式引入，代码如下：

```
<script src = "https://cdn.jsdelivr.net/npm/vue/dist/vue.js"></script>
```

3. Vue-cli 脚手架

利用 Vue-cli 脚手架构建 Vue 项目，在第 7 章将详细讲解。在中大型项目中推荐使用。

2.3　Vue 开发工具

前端开发工具有很多，笔者偏爱 WebStorm，WebStorm 是 Jetbrains 公司旗下一款 JavaScript 开发工具。已经被广大中国 JS 开发者誉为"Web 前端开发神器""最强大的 HTML5 编辑器""最智能的 JavaScript IDE"等。其与 IntelliJ IDEA 同源，继承了 IntelliJ IDEA 强大的 JS 部分功能。

1. 下载 WebStorm 工具

官网地址：https://www.jetbrains.com/webstorm/download/#section=Windows。WebStorm 的安装也比较简单，基本一路单击"下一步"按钮就可以了。

注意：WebStorm 本身不是免费的，应通过正规渠道进行安装。

2. 安装 WebStorm

安装成功后，就可以创建第一个项目了，如图 2-2 所示。

3. JavaScript 版本修改为最新的 ES6

在学习 Vue.js 之前，首先要了解 ES6。ECMAScript 6（简称 ES6）是 JavaScript 的语言标准。Vue.js 需要 ES6 的语法支持，所以需要更改 WebStorm 中的 ECMAScript 版本，如图 2-3 所示。

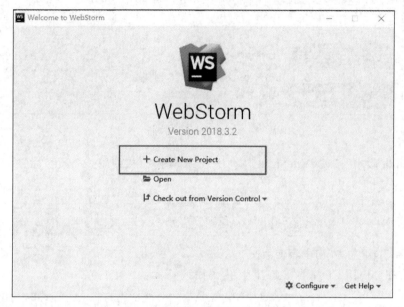

图 2-2　WebStorm 安装成功界面

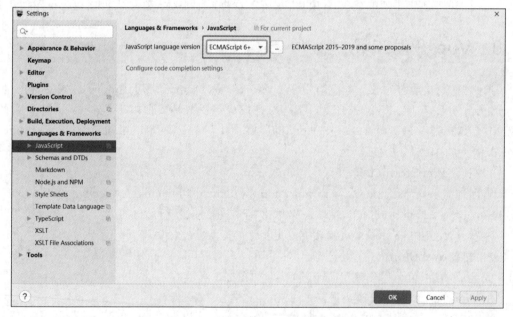

图 2-3　更改 JavaScript 版本为 ES6

2.4 第一个 Vue 程序

在使用 Vue 的时候需要通过构造函数 Vue() 创建一个 Vue 根实例，这个根实例是 View 和 Model 交互的桥梁，格式代码如下：

```
var App = new Vue({
    //编写选项代码
});
```

使用关键字 new 调用 Vue 构造器创建一个新的 Vue 实例，在实例中需要传入一些选项对象，选项对象包括 el(挂载元素)、data(数据)、methods(方法)、template(模板)、生命周期钩子函数等。例 2-1 演示创建 Vue 实例的过程。

【例 2-1】 第一个 Vue 程序

```
//第 2 章/first.html
<!DOCTYPE html>
<html>
<head lang="en">
<meta charset="UTF-8">
<title></title>
<!-- 使用 CDN 方式引入 Vue.js -->
<script src="https://cdn.jsdelivr.net/npm/vue/dist/vue.js"></script>
</head>
<body>
<!-- 创建了一个 id 为 App 的 div 标签 -->
<div id="app">
<!-- 用来输出 Vue 对象中 message 的值。如果 message 内容改变,这里的输出也会改变。-->
    {{message}}
</div>
<script>
/* 在 script 标签内,创建了 Vue 实例对象,该对象有两个属性: el 和 data。el 属性的作用是将 Vue 实例绑定到 id 为 App 的 DOM 中,data 属性用于数据的存储 */
    var app = new Vue({
        el:'#app',
        data:{
            message:'Hello world'
        }
    })
</script>
</body>
</html>
```

在 Chrome 浏览器上运行第一个 Vue 应用程序，并按 F12 键进入开发者工具模式，如

图 2-4 所示。

图 2-4　Vue 运行结果

在控制台输入 app.message='Hello Vue',然后按 Enter 键,浏览器中显示的内容会直接变成 Hello Vue,如图 2-5 所示。

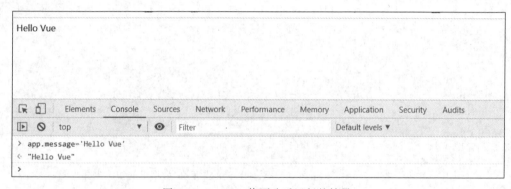

图 2-5　message 值更改后运行的结果

说明：此时可以在控制台直接输入 app.message 修改 message 的值,中间可以省略 data,在这个操作中,笔者并没有主动操作 DOM,只是让页面的内容发生了变化,这就是通过 Vue 的数据绑定功能实现的。在 MVVM 模式中 Vue 要求 ViewModel 层使用观察者模式实现数据的监听与绑定,以此做到数据与视图的快速响应。

第 3 章 Vue 基础语法

前端框架很多,无论选择哪个框架,对于前端处理的事情依旧是模板渲染、处理用户数据交互、事件的绑定等,只不过每个框架的理念和写法不同而已。Vue 通过声明一个唯一的页面实例 new Vue({…})来标记当前页面的 HTML 结构、数据展示及相关事件的绑定。

3.1 模板语法

Vue.js 使用了基于 HTML 的模板语法,允许开发者声明式地将 DOM 与底层 Vue 实例的数据进行绑定。所有 Vue.js 的模板都是合法的 HTML,所以能被遵循规范的浏览器和 HTML 解析器解析。

在底层的实现上,Vue 将模板编译成虚拟 DOM 渲染函数。结合响应系统,Vue 能够智能地计算出最少需要重新渲染多少组件,并把 DOM 操作次数减到最少。

简单地说:Vue 模板语法实现前端渲染,前端渲染即是把数据填充到 html 标签中。数据(来自服务器)+模板(html 标签)=前端渲染(产物是静态 html 内容)。

3.1.1 插值

插值就是将数据插入 html 文档中,包含文本、html 元素、元素属性等。

1. 文本插值

文本插值中用得最多的就是用双大括号的形式插入文本,如例 3-1 所示。

【例 3-1】 文本插值方式

```
//第3章/文本插值.html
<!DOCTYPE html>
<html>
<head lang="en">
<meta charset="UTF-8">
<title></title>
<script src="https://cdn.jsdelivr.net/npm/vue/dist/vue.js"></script>
</head>
```

```html
<body>
<div id="app">
    <!-- 与实例中【data】的属性绑定在一起,并且数据实现同步 -->
    <h1>{{message}}</h1>
</div></body>
<script>
    // Vue 所做的工作就是把数据填充到页面的标签中
    var vm = new Vue({
        el: "#app",
        // data 模型数据,其值是一个对象
        data: {
            message: "I LOVE YOU"
        }
    })
</script>
</body>
</html>
```

在上面代码中,data 的值更新之后不需要操作 html,页面会自动更新数据。

也可以只绑定数据一次,以后在更新 data 的属性时不需要再更新页面数据,如例 3-2 所示。

【例 3-2】 使用 v-once 指令插值

```html
//第 3 章/使用 v-once 指令插值.Html
<!DOCTYPE html>
<html>
<head lang="en">
<meta charset="UTF-8">
<title></title>
<script src="https://cdn.jsdelivr.net/npm/vue/dist/vue.js"></script>
</head>
<body>
<div id="app">
<!-- v-once 只编译一次。显示内容之后不再具有响应功能 -->
<!-- v-once 的应用场景,如果显示信息后不需要再修改,可以使用 v-once 指令,这样可以提高
性能,因为 Vue 就不需要去监听它的变化了 -->
    <h1 v-once>{{message}}</h1>
</div></body>
<script>
    // Vue 所做的工作就是把数据填充到页面的标签里
    var vm = new Vue({
        el: "#app",
        // data 模型数据,其值是一个对象
        data: {
```

```
            message: "I LOVE YOU"
        }
    })
</script>
</body>
</html>
```

在上面代码中页面只会呈现 I LOVE YOU,当改变 data 中的 message 属性值时,页面将不再刷新。

2. HTML 插值

双大括号会将数据解释为普通文本,而非 HTML 代码。为了输出真正的 HTML,需要使用 v-html,如例 3-3 所示。

【例 3-3】 使用 v-html 指令将数据解释成 HTML 代码

```
//第3章/v-html指令.html
<!DOCTYPE html>
<html>
<head lang="en">
<meta charset="UTF-8">
<title></title>
<script src="https://cdn.jsdelivr.net/npm/vue/dist/vue.js"></script>
</head>
<body>
<div id="app">
<!-- 内容按普通 HTML 插入,不会作为 Vue 模板进行编译,在网站上动态渲染任意 HTML 是非常危险的,因为容易导致 XSS 攻击 -->
    <h1 v-html="msg"></h1>
</div>
</body>
<script>
    var vm = new Vue({
        el: "#app",
        data: {
            msg: "<span style='color:blue'>BLUE</span>"/* 可以使用 v-html 标签展示 HTML 代码。*/
        }
    })
</script>
</html>
```

上面代码将 msg 属性值作为 html 元素插入 h1 标签的子节点中。

3. 属性插值

Mustache 语法不能作用在 HTML attribute 上,遇到这种情况应该使用 v-bind 指令。

在开发的时候，有时属性不是固定不变的，有可能根据我们的一些数据动态地决定，例如图片标签(<img>)的src属性，我们可能从后端请求了一个包含图片地址的数组，需要将地址动态地绑定到src上，这时就不能简单地将src固定为某个地址。还有一个例子就是a标签的href属性。这时可以使用v-bind:指令。其中，v-bind:可以缩写成冒号:，如例3-4所示。

【例3-4】 使用v-bind:指令绑定属性

```html
//第3章/使用v-bind指令.html
<!DOCTYPE html>
<html xmlns:v-bind="http://www.w3.org/1999/xhtml">
<head lang="en">
<meta charset="UTF-8">
<title></title>
<script src="https://cdn.jsdelivr.net/npm/vue/dist/vue.js"></script>
</head>
<body>
<div id="app">
    <img v-bind:src="imgUrl" alt=""/>
    <a :href="searchUrl">百度一下</a>
</div>
</body>
<script>
    var vm = new Vue({
        el: "#app",
        data: {
            imgUrl:'images/1.jpg',
            searchUrl:'http://www.baidu.com'
        }
    })
</script>
</html>
```

服务器请求过来的数据，我们一般都会在data那里中转一下，中转过后再把需要的变量绑定到对应的属性上面。

v-bind除了可以在开发中用在有特殊意义的属性外(src、href等)，也可以绑定其他一些属性，如与class和style绑定，如例3-5和3-6所示。

【例3-5】 v-bind动态绑定属性class

```html
//第3章/v-bind动态绑定属性.html
<!DOCTYPE html>
<html lang="en" xmlns:v-bind="http://www.w3.org/1999/xhtml">
<head>
<meta charset="UTF-8">
<title>v-bind动态绑定属性class</title>
<script src="https://cdn.jsdelivr.net/npm/vue/dist/vue.js"></script>
```

```
</head>
<body>
<div id="app">
<p :class="{fontCol:isName,setBack:!isAge}" class="weight">{{name}}</p>
<i :class="addClass">{{name}}真好看!</i>
</div>
<script>
    var vm = new Vue({
        el:"#app",
        // 条件比较少
        data:{
            isName:true,
            isAge:false,
            name:"功守道"
        },
        /* 当v-bind:class的表达式过长或者逻辑复杂时(一般当条件多于两个的时候),可以考虑
        采用计算属性,返回一个对象 */
        computed:{
            addClass:function(){
                return {
                    checked:this.isName&&!this.isAge
                }
            }
        }
    })
</script>
</body>
</html>
```

既然是一个对象,那么该对象内的属性就可能不唯一,但只要有一项为真,对应的类名就会存在。通过v-bind更新的类名和元素本身存在的类名不冲突,可以优雅地共存,如图3-1所示。

图3-1　v-bind动态绑定属性class

【例3-6】 v-bind 动态绑定属性 style

```
//第3章/v-bind动态绑定属性style.html
<!DOCTYPE html>
<html lang="en" xmlns:v-bind="http://www.w3.org/1999/xhtml">
<head>
<meta charset="UTF-8">
<title>v-bind动态绑定属性style</title>
<script src="https://cdn.jsdelivr.net/npm/vue/dist/vue.js"></script>
</head>
<body>
<div id="app">
<div :style="{'color': color,'fontSize':fontSize + 'px'}">修饰文本</div>
</div>
<script>
    var vm = new Vue({
        el: "#app",
        data: {
            color: 'red',
            fontSize: 24
        }
    })
</script>
</body>
</html>
```

在大多情况下，标签中直接写一长串属性的样式不便于阅读和维护，所以一般写在 dada 或 computed 里，代码如下：

```
<div id="app">
<div :style="styles">修饰文本</div>
</div>
<script>
Var vm = new Vue({
    el: "#app",
    data: {
        styles:{
          color: 'red',
          fontSize: 14 + 'px'
        }
    }
})
</script>
```

应用多个样式对象时,可以使用数组语法,如例 3-7 所示。

【例 3-7】 v-bind 绑定多个 style 属性

```
//第 3 章/v-bind 绑定多个 style 属性.html
<!DOCTYPE html>
<html lang="en" xmlns:v-bind="http://www.w3.org/1999/xhtml">
<head>
<meta charset="UTF-8">
<title>v-bind 动态绑定属性 style</title>
<script src="https://cdn.jsdelivr.net/npm/vue/dist/vue.js"></script>
</head>
<body>
<div id="app">
    <div :style="[styleA,styleB]">文本</div>
</div>
<script>
    var vm = new Vue({
        el: "#app",
        data: {
            styleA:{
                color: 'red',
                fontSize: 24 + 'px'
            },
            styleB: {
                width: 100 + 'px',
                border: 1 + 'px ' + 'black ' + 'solid'
            }
        }
    })
</script>
</body>
</html>
```

4. 插值中使用 JavaScript 表达式

迄今为止,在模板中,我们一直都只绑定简单的 property 键值。但实际上,对于所有的数据绑定,Vue.js 都提供了完全的 JavaScript 表达式支持,如例 3-8 所示。部分表达式格式代码如下:

```
{{ number + 1 }}
{{ ok ? 'YES' : 'NO' }}
{{ message.split('').reverse().join('') }}
<div v-bind:id="'list-' + id"></div>
```

这些表达式会在所属 Vue 实例的数据作用域下作为 JavaScript 被解析。有个限制需要注意，每个绑定只能包含单个表达式，所以下面的表达式不会生效，代码如下：

```
<!-- 这是语句,不是表达式 -->
{{ var a = 1 }}
<!-- 流控制不会生效,请使用三元表达式 -->
{{ if (ok) { return message } }}
```

【例 3-8】 使用 JavaScript 表达式

```
//第 3 章/使用 JavaScript 表达式.html
<!DOCTYPE html>
<html lang="en" xmlns:v-bind="http://www.w3.org/1999/xhtml">
<head>
<meta charset="UTF-8">
<title></title>
<script src="https://cdn.jsdelivr.net/npm/vue/dist/vue.js"></script>
</head>
<body>
<div id="app">
    <p>{{ number + 1 }}</p>
    <hr>
    <p>{{msg + '~~~~~'}}</p>
    <hr>
    <p>{{flag ? '条件为真' : '条件为假'}}</p>
    <hr>
</div>
<script>
    var vm = new Vue({
        el:'#app',
        data:{
            msg:'Hello beixi!',
            flag:true,
            number:2
        }
    })
</script>
</body>
</html>
```

3.1.2 指令

指令在上面已经使用过了,例如 v-bind:和 v-html,指令就是指这些带有 v-前缀的特殊属性。指令属性的值预期是单一 JavaScript 表达式(除了 v-for)。指令的职责就是当其表达式的值改变时相应地将某些行为应用到 DOM 上。

1. 参数

有一些指令能够接收一个"参数",在指令名称之后以冒号表示。如 v-bind 指令可以用于响应式地更新 HTML attribute,代码如下:

```
<a v-bind:href = "url">...</a>
```

在这里 href 是参数,告知 v-bind 指令将该元素的 href attribute 与表达式 url 的值绑定。

2. 动态参数

从版本 2.6.0 开始,可以用方括号将 JavaScript 表达式作为一个指令的参数括起来,响应式使得 Vue 更加灵活多变,其动态参数也有其含义,代码如下:

```
<a v-bind:[attributeName] = 'url'>...</a>
```

这里的 attributeName 会被作为一个 JavaScript 表达式进行动态求值,求得的值将作为最终的参数使用。例如,如果 Vue 实例有一个 data 属性 attributeName,其值为'href',那么这个绑定将等价于 v-bind:href。

同样地,可以使用动态参数作为一个动态的事件名绑定并处理函数,格式代码如下:

```
<a v-on:[eventName] = 'doSomething'>...</a>
```

同样地,当 eventName 的值为'focus'时,v-on:[eventName]等价于 v-on:focus。

当然动态参数的值也有约束,动态参数预期会求出一个字符串,异常情况下此值为 null。这个特殊的 null 值可以被显性地移出绑定。任何其他非字符串类型的值都会触发一个警告。

3. 修饰符

修饰符(Modifiers)是以半角句号"."指明的特殊后缀,用于指出一个指令应该以特殊方式绑定。例如,.prevent 修饰符告诉 v-on 指令对于触发的事件调用 event.preventDefault(),代码如下:

```
<form v-on:submit.prevent = "onSubmit">...</form>
```

在后面章节对 v-on 和 v-for 等功能的探索中,将会看到修饰符的其他例子。

3.1.3 过滤器

过滤器是对即将显示的数据做进一步的筛选和处理,然后进行显示,值得注意的是过滤器并没有改变原来的数据,只是在原数据的基础上产生新的数据。

过滤器分全局过滤器和局部过滤器,全局过滤器在项目中使用频率非常高。

1. 定义过滤器

- 全局过滤器

```
Vue.filter('过滤器名称', function (value1[,value2,...] ) {
//逻辑代码
})
```

- 局部过滤器

```
new Vue({
        filters: {
       '过滤器名称': function (value1[,value2,...] ) {
          // 逻辑代码
            }
      }
})
```

2. 过滤器使用的地方

Vue.js 允许自定义过滤器,可被用于一些常见的文本格式化。过滤器可以用在两个地方:双花括号插值和 v-bind 表达式(后者从 2.1.0＋版开始支持)。过滤器应该被添加在 JavaScript 表达式的尾部,由"管道"符号指示,格式代码如下:

```
<!-- 在双花括号中 -->
<div>{{数据属性名称 | 过滤器名称}}</div>
<div>{{数据属性名称 | 过滤器名称(参数值)}}</div>
<!-- 在 v-bind 中 -->
<div v-bind:id = "数据属性名称 | 过滤器名称"></div>
<div v-bind:id = "数据属性名称 | 过滤器名称(参数值)"></div>
```

3. 实例

局部过滤器在实例中的使用如例 3-9 所示。

【例 3-9】 在价格前面加上人民币符号(￥)

```
//第 3 章/局部过滤器.html
<!DOCTYPE html>
```

```
<html lang="en" xmlns:v-bind="http://www.w3.org/1999/xhtml">
<head>
<meta charset="UTF-8">
<title></title>
<script src="https://cdn.jsdelivr.net/npm/vue/dist/vue.js"></script>
</head>
<body>
<div id="app">
<!-- 文本后边需要添加管道符号(|)作为分隔,管道符"|"后边是文本的处理函数,处理函数的第
一个参数是管道符前边的文本内容,如果处理函数用于传递参数,则从第二个参数往后依次是传递
的参数 -->
    <p>计算机价格：{{price | addPriceIcon}}</p>
</div>
<script>
    var vm = new Vue({
        el:"#app",
        data:{
            price:200
        },
        filters:{
            //处理函数
            addPriceIcon(value){
                console.log(value)//200
                return '¥' + value;
            }
        }
    })
</script>
</body>
</html>
```

传递多个参数的局部过滤器如例3-10所示。

【例3-10】 在局部过滤器中传递多个参数

```
//第3章/在局部过滤器中传递多个参数.html
<!DOCTYPE html>
<html lang="en" xmlns:v-bind="http://www.w3.org/1999/xhtml">
<head>
<meta charset="UTF-8">
<title></title>
<script src="https://cdn.jsdelivr.net/npm/vue/dist/vue.js"></script>
</head>
<body>
<div id="app">
```

```html
            <!-- 过滤器接收多个参数 -->
            <span>{{value1 | multiple(value2,value3)}}</span>
    </div>
    <script>
        var vm = new Vue({
            el: '#app',
            data: {
                msg: 'hello',
                value1:10,
                value2:20,
                value3:30
            },
            //局部过滤器
            filters: {
                'multiple': function (value1, value2, value3) {
                    return value1 * value2 * value3
                }
            }
        })
    </script>
</body>
</html>
```

全局过滤器在实例中的使用如例 3-11 所示。

【例 3-11】 全局过滤器的使用

```html
//第 3 章/全局过滤器.html
<!DOCTYPE html>
<html lang="en" xmlns:v-bind="http://www.w3.org/1999/xhtml">
<head>
<meta charset="UTF-8">
<title></title>
<script src="https://cdn.jsdelivr.net/npm/vue/dist/vue.js"></script>
</head>
<body>
<div id="app">
    <h3>{{viewContent | addNamePrefix}}</h3>
</div>
<script>
    /* addNamePrefix 是过滤器的名字,也是管道符后边的处理函数; value 是参数 */
    Vue.filter("addNamePrefix",(value) =>{
        return "my name is " + value
    })
    var vm = new Vue({
```

```
            el:"#app",
            data:{
                viewContent:"贝西"
            }
        })
    </script>
</body>
</html>
```

3.2 实例及选项

Vue 通过构造函数来实例化一个 Vue 对象：var vm＝new Vue({})。在实例化时，我们会传入一些选项对象，包含数据选项、属性选项、方法选项、生命周期钩子等常用选项。而且 Vue 的核心是一个响应式的数据绑定系统，建立绑定后，DOM 将和数据保持同步，这样无须手动维护 DOM，就能使代码更加简洁易懂，从而提升效率。

3.2.1 数据选项

一般地，当模板内容较简单时，使用 data 选项配合表达式即可。当涉及复杂逻辑时，则需要用到 methods、computed、watch 等方法。

data 是 Vue 实例的数据对象。Vue 使用递归法将 data 的属性转换为 getter/setter，从而让 data 属性能够响应数据变化，代码如下：

```
<!-- 部分代码省略 -->
<div id="app">
  {{ message }}
</div>
<script>
var values = {message: 'Hello Vue!'}
var vm = new Vue({
  el: '#app',
  data: values
})
console.log(vm);
</script>
```

Vue 实例创建之后，在控制台输入 vm.$data 即可访问原始数据对象，如图 3-2 所示。

在 script 标签中添加一些输出信息即可查看控制台从而观察 Vue 实例是否代理了 data 对象的所有属性，代码如下：

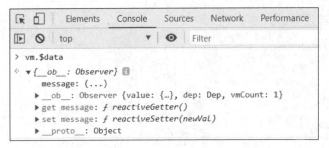

图 3-2　访问原始数据对象

```
<script>
  console.log(vm.$data === values);        //true
  console.log(vm.message);                 //'Hello Vue!'
  console.log(vm.$data.message);           //'Hello Vue!'
</script>
```

被代理的属性是响应式的,也就是说值的任何改变都将触发视图的重新渲染。设置属性也会影响到原始数据,反之亦然。

但是,以"_"或"$"开头的属性不会被 Vue 实例代理,因为它们可能和 Vue 内置的属性或方法冲突。可以使用如 vm.$data._property 的方式访问这些属性,代码如下:

```
<script>
var values = {
  message: 'Hello Vue!',
  _name: 'beixi'
}
var vm = new Vue({
  el: '#app',
  data: values
})
console.log(vm._name);                   //undefined
console.log(vm.$data._name);             //'beixi'
</script>
```

3.2.2　属性选项

Vue 为组件开发提供了属性(props)选项,可以使用它为组件注册动态属性,来处理业务之间的差异性,使代码可以在相似的应用场景复用。

props 选项可以是数组或者对象类型,用于接收从父组件传递过来的参数,并允许为其赋默认值、类型检查和规则校验等,如例 3-12 所示。

【例 3-12】 props 选项的使用

```html
//第 3 章/props 选项的使用.Html
<!DOCTYPE html>
<html>
<head>
<title>Hello World</title>
<script src='http://cdnjs.cloudflare.com/ajax/libs/vue/1.0.26/vue.min.js'></script>
</head>
<body>
<div id="app">
    <message content='Hello World'></message>
</div>
</body>
<!-- 测试组件 -->
<script type="text/javascript">
    var Message = Vue.extend({
        props : ['content'],
        data : function(){
            return {
                a: 'it worked'
            }},
        template : '<h1>{{content}}</h1><h1>{{a}}</h1>'
    })
    Vue.component('message', Message)
    var vm = new Vue({
        el : '#app',
    })
</script>
</html>
```

3.2.3 方法选项

可以通过选项属性 methods 对象来定义方法，并且使用 v-on 指令来监听 DOM 事件，如例 3-13 所示。

【例 3-13】 通过 methods 对象来定义方法

```html
//第 3 章/通过 methods 对象来定义方法.html
<!DOCTYPE html>
<html xmlns:v-on="http://www.w3.org/1999/xhtml">
<head>
<title></title>
<meta charset="utf-8"/>
<script src='http://cdnjs.cloudflare.com/ajax/libs/vue/1.0.26/vue.min.js'></script>
```

```html
</head>
<body>
<div id="app">
<button v-on:click="test">点我</button>
</div>
</body>
<!-- 测试组件 -->
<script type="text/javascript">
    var vm = new Vue({
        el: '#app',
        methods:{
/*定义了一个test函数*/
            test: function () {
                console.log(new Date().toLocaleTimeString());
            }
        }
    })
</script>
</html>
```

3.2.4 计算属性

在项目开发中,我们展示的数据往往需要处理。除了在模板中绑定表达式或利用过滤器外,Vue 还提供了计算属性方法,计算属性是当其依赖属性的值发生变化时,这个属性的值会自动更新,与之相关的 DOM 部分也会同步自动更新,从而减轻在模板中的业务负担,保证模板的结构清晰和可维护性。

有时候需要在{{}}中进行一些计算再展示出数据,如例 3-14 所示。

【例 3-14】 在页面中展示学生的成绩、总分和平均分

```html
//第3章/学生成绩信息.html
<!DOCTYPE html>
<html xmlns:v-on="http://www.w3.org/1999/xhtml">
<head>
<title></title>
<meta charset="utf-8"/>
<script src="https://cdn.jsdelivr.net/npm/vue/dist/vue.js"></script>
</head>
<body>
<div id="app">
<table border="1">
<thead>
    <th>学科</th>
    <th>分数</th>
```

```
        </thead>
        <tbody>
        <tr>
            <td>数学</td>
            <td><input type="text" v-model="Math"></td>
        </tr>
        <tr>
            <td>英语</td>
            <td><input type="text" v-model="English"></td>
        </tr>
        <tr>
            <td>语文</td>
            <td><input type="text" v-model="Chinese"></td>
        </tr>
        <tr>
            <td>总分</td>
            <td>{{Math + English + Chinese}}</td>
        </tr>
        <tr>
            <td>平均分</td>
            <td>{{(Math + English + Chinese)/3}}</td>
        </tr>
        </tbody>
        </table>
        <script>
            var vm = new Vue({
                el:'#app',
                data:{
                    Math:66,
                    English:77,
                    Chinese:88
                }
            })
        </script>
        </div>
        </body>
        </html>
```

执行结果如图3-3所示。

虽然通过{{ }}运算,可以解决我们的问题,但是代码结构不清晰,特别是当运算比较复杂的时候,我们不能复用功能代码。这时,大家不难想到用methods来封装功能代码,但事实上,Vue提供了一个更好的解决方案——计算属性。计算属性是Vue实例中的一个配置选项: computed,通常计算相关的函数,返回最后计算出来的值。也就是我们可以把这些计算的过

学科	分数
数学	66
英语	77
语文	88
总分	231
平均分	77

图3-3 学生成绩、总分和
平均分的展示

程写到一个计算属性中,然后让它动态地计算,如例3-15所示。

【例3-15】 使用计算属性展示学生的成绩、总分和平均分

```
//第3章/使用计算属性展示学生成绩.html
<!DOCTYPE html>
<html xmlns:v-on="http://www.w3.org/1999/xhtml">
<head>
<title></title>
<meta charset="utf-8"/>
<script src="https://cdn.jsdelivr.net/npm/vue/dist/vue.js"></script>
</head>
<body>
<div id="app">
<table border="1">
<thead>
    <th>学科</th>
    <th>成绩</th>
</thead>
<tbody>
<tr>
    <td>数学</td>
    <td><input type="text" v-model.number="Math"></td>
</tr>
<tr>
    <td>英语</td>
    <td><input type="text" v-model.number="English"></td>
</tr>
<tr>
    <td>语文</td>
    <td><input type="text" v-model.number="Chinese"></td>
</tr>
<tr>
    <td>总分</td>
    <td>{{sum}}</td>
</tr>
<tr>
    <td>平均分</td>
    <td>{{average}}</td>
</tr>

</tbody>
</table>
</div>
<script>
    var vm = new Vue({
```

```
            el:'#app',
            data:{
                Math:66,
                English: 77,
                Chinese:88
            },
            computed:{
                    <!-- 一个计算属性的 getter -->
                sum:function(){
                    <!-- this 指向 vm 实例 -->
                    return this.Math + this.English + this.Chinese;
                },
                average:function(){
                    return Math.round(this.sum/3);
                }
            }
        });
    </script>
</body>
</html>
```

计算属性一般通过其他的数据计算出一个新数据,它的一个好处是能把新的数据缓存下来,当其他的依赖数据没有发生改变时,它调用的是缓存数据,这就极大地提高了程序的性能和数据提取的速度。而如果将计算过程写在 methods 中,数据就不会缓存下来,所以每次都会重新计算。这也是我们没有采用 methods 的原因,如例 3-16 所示。

【例 3-16】 计算属性和方法的比较

```
//第 3 章/计算属性和方法的比较.html
<!DOCTYPE html>
<html xmlns:v-on = "http://www.w3.org/1999/xhtml">
<head>
<title></title>
<meta charset = "utf-8"/>
<script src = "https://cdn.jsdelivr.net/npm/vue/dist/vue.js"></script>
</head>
<body>
<div id = "app">
    <p>原始字符串:"{{ message }}"</p>
    <p>计算属性反向字符串:"{{ reversedMessage1 }}"</p>
    <p>methods 方法反向字符串:"{{ reversedMessage2() }}"</p>
</div>
<script>
    var vm = new Vue({
```

```
            el: '#app',
            data: {
                message: 'beixi'
            },
            computed:{
                reversedMessage1: function () {
                    return this.message.split('').reverse().join('')
                }
            },
            methods: {
                reversedMessage2: function () {
                    return this.message.split('').reverse().join('')
                }
            }
        })
    </script>
</body>
</html>
```

计算属性是基于它们的依赖进行缓存的。计算属性只有在它的相关依赖发生改变时才会重新求值。这就意味着只要 message 没有发生改变,多次访问 reversedMessage1 计算属性会立即返回之前的计算结果,而不必再次执行函数。相比而言,只要发生重新渲染,methods 总会调用并执行该函数。

假设有一个性能开销比较大的计算属性 A,它需要遍历一个极大的数组和做大量的计算。可能有其他的计算属性依赖于 A,如果没有缓存,则将不可避免地多次执行 A 的 getter,这样便极大地降低了数据的提取速度,从而导致用户体验感不强。如果不希望有缓存,则用 methods 替代。

实例 3-15 和实例 3-16 只提供了 getter 读取值的方法,实际上除了 getter,还可以设置计算属性的 setter,如例 3-17 所示。

【例 3-17】 读取和设置值

```
//第3章/读取和设置值.html
<!DOCTYPE html>
<html xmlns:v-on="http://www.w3.org/1999/xhtml">
<head>
<title></title>
<meta charset="utf-8"/>
<script src="https://cdn.jsdelivr.net/npm/vue/dist/vue.js"></script>
</head>
<body>
<script>
    var vm = new Vue({
```

```
            data: { a: 1 },
            computed: {
                //该函数只能读取数据,使用 vm.aDouble 即可读取数据
                aDouble: function () {
                    return this.a * 2
                },
                //读取和设置数据
                aPlus: {
                    get: function () {
                        return this.a + 1
                    },
                    set: function (v) {
                        this.a = v - 1
                    }
                }
            }
        })
        console.log(vm.aPlus);            //2
        vm.aPlus = 3;
        console.log(vm.a);                //2
        console.log(vm.aDouble);          //4
    </script>
</body>
</html>
```

3.2.5　表单控件

1. 基础用法

可以用 v-model 指令在表单< input >、< textarea >及< select >元素上创建双向数据绑定,如图 3-4 所示。它会根据控件类型自动选取正确的方法来更新元素。尽管有些神奇,但 v-model 本质上不过是语法糖。它负责监听用户的输入事件以便更新数据,并对一些极端场景进行一些特殊处理。

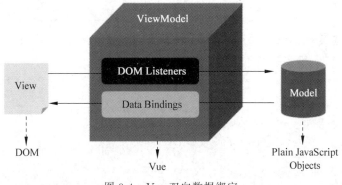

图 3-4　Vue 双向数据绑定

注意：v-model 会忽略所有表单元素的 value、checked、selected attribute 的初始值而总是将 Vue 实例的数据作为数据来源，所以应该通过 JavaScript 在组件的 data 选项中声明初始值。

（1）单行文本可在 input 元素中使用 v-model 实现双向数据绑定，代码如下：

```html
<div id="app" class="demo">
    <input v-model="message" placeholder="请输入信息...">
    <p>Message is: {{ message }}</p>
</div>
<script>
    var vm = new Vue({
        el: '#app',
        data: {
            message: ''
        }
    })
</script>
```

（2）多行文本可在 textarea 元素中使用 v-model 实现双向数据绑定，代码如下：

```html
<div id="example-textarea" class="demo">
    <span>Multiline message is:</span>
    <p style="white-space: pre">{{ message }}</p>
    <br>
    <textarea v-model="message" placeholder="add multiple lines"></textarea>
</div>
<script>
    new Vue({
        el: '#example-textarea',
        data: {
            message: ''
        }
    })
</script>
```

注意：在文本区域插值（<textarea></textarea>）并不会生效，应用 v-model 来代替。

（3）单个复选框，绑定到布尔值，代码如下：

```html
<div id="example-checkbox" class="demo">
    <input type="checkbox" id="checkbox" v-model="checked">
    <label for="checkbox">{{ checked }}</label>
</div>
<script>
    new Vue({
```

```
        el: '#example-checkbox',
        data: {
            checked: false
        }
    })
</script>
```

(4) 多个复选框,绑定到同一个数组,代码如下:

```
<div id="example-checkboxs" class="demo">
    <input type="checkbox" id="jack" value="Jack" v-model="checkedNames">
    <label for="jack">Jack</label>
    <input type="checkbox" id="john" value="John" v-model="checkedNames">
    <label for="john">John</label>
    <input type="checkbox" id="mike" value="Mike" v-model="checkedNames">
    <label for="mike">Mike</label>
    <br>
    <span>Checked names: {{ checkedNames }}</span>
</div>
<script>
    new Vue({
        el: '#example-checkboxs',
        data: {
            checkedNames: []
        }
    })
</script>
```

(5) 单选按钮的双向数据绑定,代码如下:

```
<div id="example-radio">
    <input type="radio" id="runoob" value="Runoob" v-model="picked">
    <label for="runoob">Runoob</label>
    <br>
    <input type="radio" id="google" value="Google" v-model="picked">
    <label for="google">Google</label>
    <br>
    <span>选中值为: {{ picked }}</span>
</div>
<script>
    new Vue({
        el: '#example-radio',
        data: {
            picked : 'Runoob'
```

```
        }
    })
</script>
```

(6) 选择列表、下拉列表的双向数据绑定。

- 单选时，代码如下：

```
<div id="example-selected">
<select v-model="selected">
    <option disabled value="">请选择</option>
    <option>A</option>
    <option>B</option>
    <option>C</option>
</select>
<span>Selected: {{ selected }}</span>
</div>
<script>
    new Vue({
        el: '#example-selected',
        data: {
            selected: ''
        }
    })
</script>
```

注意：如果 v-model 表达式的初始值未能匹配任何选项，则<select>元素将被渲染为"未选中"状态。在 iOS 中，这会使用户无法选择第一个选项。因为在这种情况下，iOS 不会触发 change 事件。因此，更推荐像上面那样提供一个值为空的禁用选项。

- 多选列表(绑定到一个数组)，代码如下：

```
<div id="example-selected" class="demo">
    <select v-model="selected" multiple style="width: 50px">
        <option>A</option>
        <option>B</option>
        <option>C</option>
    </select>
    <br>
    <span>Selected: {{ selected }}</span>
</div>
<script>
    new Vue({
        el: '#example-selected',
        data: {
```

```
            selected: []
        }
    })
</script>
```

- 动态选项,用 v-for 渲染,代码如下:

```
<div id = "example - selected" class = "demo">
    <select v - model = "selected">
        <option v - for = "option in options" v - bind:value = "option.value">
            {{ option.text }}
        </option>
    </select>
<span> Selected: {{ selected }}</span>
</div>
<script>
    new Vue({
        el: '#example - selected',
        data: {
            selected: 'A',
            options: [
                { text: 'One', value: 'A' },
                { text: 'Two', value: 'B' },
                { text: 'Three', value: 'C' }
            ]
        }
    })
</script>
```

2. 绑定 value

(1) 对于单选按钮、勾选框及选择列表选项,v-model 绑定的 value 通常是静态字符串,代码如下:

```
<div id = "app">
    <!-- 当选中时,picked 为字符串 "a" -->
    <input type = "radio" v - model = "picked" value = "a">
    <!-- 当选中时,toggle 为 true 或 false -->
    <input type = "checkbox" v - model = "toggle">
    <!-- 当选中时,selected 为字符串 "abc" -->
    <select v - model = "selected">
        <option value = "abc">ABC</option>
    </select>
</div>
<script>
```

```
var vm = new Vue({
    el:'#app',
    data:{
        picked:'',
        toggle:'',
        selected:''
    }
})
</script>
```

（2）实现复选框功能。有时我们想绑定value到Vue实例的一个动态属性上，这时可以用v-bind实现，并且这个属性的值可以不是字符串，代码如下：

```
<div id="app">
<input type="checkbox" v-model="toggle" v-bind:true-value="a" v-bind:false-value="b">
    {{toggle}}
</div>
<script>
var vm = new Vue({
    el:'#app',
    data:{
        toggle:'',
        a:'a',
        b:'b'
    }
})
</script>
```

当选中时输出字符串a；当没有选中时输出字符串b。

（3）单选按钮，代码如下：

```
<input type="radio" v-model="pick" v-bind:value="a">
```

当选中时在页面控制台输入vm.pick的值和vm.a的值相等。

（4）选列表设置，代码如下：

```
<select v-model="selected">
    <!-- 内联对象字面量 -->
    <option v-bind:value="{ number: 123 }">123</option>
</select>
```

当选中时在页面控制台输入typeof vm.selected，则输出'object'；如果输入vm.selected.number，则输出值为123。

3. 修饰符

- .lazy

在默认情况下，v-model 每次在 input 事件触发后将输入框的值与数据进行同步（除了上述输入法组合文字时）。此时可以添加 .lazy 修饰符，从而转为在 change 事件之后进行同步，代码如下：

```html
<!-- 在"change"时而非"input"时更新 -->
<input v-model.lazy="msg">
```

- .number

如果想自动将用户的输入值转为数值类型，可以给 v-model 添加 .number 修饰符，代码如下：

```html
<input v-model.number="age" type="number">
```

通常这很有用，因为即使在 type="number" 时，HTML 输入元素的值也总会返回字符串。如果这个值无法被 parseFloat() 解析，则会返回原始的值。

- .trim

如果要自动过滤用户输入的首尾空白字符，可以给 v-model 添加 .trim 修饰符，代码如下：

```html
<input v-model.trim="msg">
```

3.2.6 生命周期

Vue 实例有一个完整的生命周期，也就是开始创建、初始化数据、编译模板、挂载 DOM、渲染→更新→渲染、卸载等一系列过程，我们称之为 Vue 的生命周期。通俗地说就是 Vue 实例从创建到销毁的过程，如图 3-5 所示。

可以看到在 Vue 的整个生命周期中有很多钩子函数，这就给了用户在不同阶段添加自己代码的机会，下面先列出所有的钩子函数，然后再一一详解。

(1) beforeCreate：在实例初始化之后，数据观测(data observer)和 event/watcher 事件配置之前被调用。

(2) Created：实例已经创建完成之后被调用。在这一步，实例已完成以下配置，数据观测(data observer)、属性和方法的运算，以及 watch/event 事件回调。然而，挂载阶段还没开始，$el 属性目前不可见。

(3) beforeMount：在挂载开始之前被调用，相关的 render 函数首次被调用。

(4) Mounted：el 被新创建的 vm.$el 替换，并挂载到实例上之后调用该钩子。

(5) beforeUpdate：数据更新时调用，发生在虚拟 DOM 重新渲染和打补丁之前。可以在这个钩子中进一步更改状态，这不会触发附加的重新渲染过程。

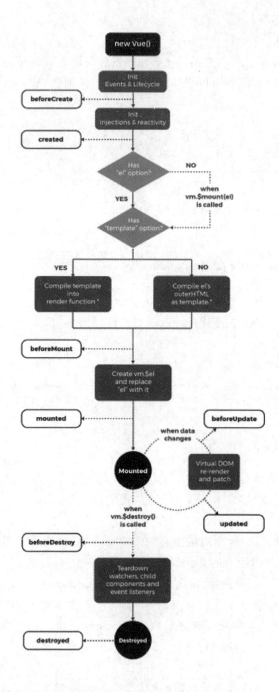

图 3-5　Vue 生命周期

（6）Updated：由于数据更改导致的虚拟 DOM 重新渲染和打补丁，在这之后会调用该钩子。当这个钩子被调用时，组件 DOM 已经更新，所以现在可以执行依赖于 DOM 的操作。然而在大多数情况下，应该避免在此期间更改状态，因为这可能会导致更新无限循环。该钩子在服务器端渲染期间不会被调用。

（7）beforeDestroy：实例销毁之前调用。在这一步，实例仍然完全可用。

（8）Destroyed：Vue 实例销毁后调用。调用后，Vue 实例指示的所有东西都会被解绑定，所有的事件监听器会被移除，所有的子实例也会被销毁。该钩子在服务器端渲染期间不会被调用。

下面通过实例代码来演示 Vue 实例在创建过程中调用的几个生命周期钩子，如例 3-18 所示。

【例 3-18】 生命周期钩子函数的演示

```html
//第3章/生命周期钩子函数.html
<!DOCTYPE html>
<html xmlns:v-on="http://www.w3.org/1999/xhtml"
xmlns:v-bind="http://www.w3.org/1999/xhtml">
<head>
<title></title>
<meta charset="utf-8"/>
<script src="https://cdn.jsdelivr.net/npm/vue/dist/vue.js"></script>
</head>
<body>
<div id="app">
<h1>{{message}}</h1>
</div>
</body>
<script>
    var vm = new Vue({
        el: '#app',
        data: {
            message: 'Vue 的生命周期'
        },
        beforeCreate: function() {
            console.group('------beforeCreate 创建前状态------');
            console.log("%c%s", "color:red", "el : " + this.$el);        //undefined
            console.log("%c%s", "color:red","data : " + this.$data);      //undefined
            console.log("%c%s", "color:red","message: " + this.message)
        },
        created: function() {
            console.group('------created 创建完毕状态------');
            console.log("%c%s", "color:red","el : " + this.$el);         //undefined
            console.log("%c%s", "color:red","data : " + this.$data);     //已被初始化
```

```js
        console.log("%c%s","color:red","message: " + this.message);    //已被初始化
},
beforeMount: function() {
    console.group('------beforeMount 挂载前状态------');
    console.log("%c%s","color:red","el : " + (this.$el));              //已被初始化
    console.log(this.$el);
    console.log("%c%s","color:red","data : " + this.$data);            //已被初始化
    console.log("%c%s","color:red","message: " + this.message);        //已被初始化
},
mounted: function() {
    console.group('------mounted 挂载结束状态------');
    console.log("%c%s","color:red","el : " + this.$el);                //已被初始化
    console.log(this.$el);
    console.log("%c%s","color:red","data : " + this.$data);            //已被初始化
    console.log("%c%s","color:red","message: " + this.message);        //已被初始化
},
beforeUpdate: function () {
    console.group('beforeUpdate 更新前状态===============»');
    console.log("%c%s","color:red","el : " + this.$el);
    console.log(this.$el);
    console.log("%c%s","color:red","data : " + this.$data);
    console.log("%c%s","color:red","message: " + this.message);
},
updated: function () {
    console.group('updated 更新完成状态===============»');
    console.log("%c%s","color:red","el : " + this.$el);
    console.log(this.$el);
    console.log("%c%s","color:red","data : " + this.$data);
    console.log("%c%s","color:red","message: " + this.message);
},
beforeDestroy: function () {
    console.group('beforeDestroy 销毁前状态===============»');
    console.log("%c%s","color:red","el : " + this.$el);
    console.log(this.$el);
    console.log("%c%s","color:red","data : " + this.$data);
    console.log("%c%s","color:red","message: " + this.message);
},
destroyed: function () {
    console.group('destroyed 销毁完成状态===============»');
    console.log("%c%s","color:red","el : " + this.$el);
```

```
                console.log(this.$el);
                console.log("%c%s","color:red","data : " + this.$data);
                console.log("%c%s","color:red","message: " + this.message)
            }
        })
    </script>
</html>
```

运行后打开console可以看到打印出来的内容,如图3-6所示。

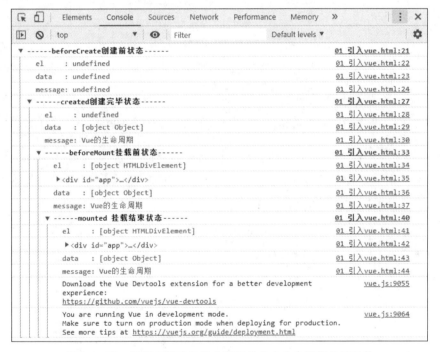

图3-6　Vue实例在创建过程中调用的几个生命周期钩子函数

接下来将详细解释生命周期钩子函数。

(1) 在beforeCreate和created钩子函数之间的生命周期。

在这个生命周期之间,首先进行初始化事件,然后进行数据观测,此时可以看到在created的时候数据已经和data属性进行了绑定(放在data中的属性的值发生改变时,视图也会改变)。

注意：此时还是没有el选项。

(2) created钩子函数和beforeMount间的生命周期,如图3-7所示。

在这一阶段发生的事情还是比较多的。首先会判断对象是否有el选项。如果有就继续向下编译,如果没有el选项,则停止编译,也就意味着停止了生命周期,直到在该Vue实

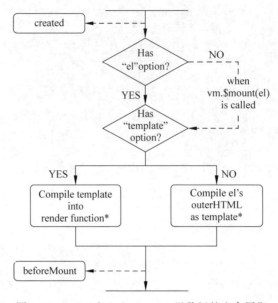

图 3-7　created 和 beforeMount 函数间的生命周期

例上调用 vm.$mount(el)。此时注释掉代码中的：el：'♯app'，然后可以看到程序运行到 created 时就停止了，如图 3-8 所示。

图 3-8　没有 el 选项的生命周期

如果在后面继续调用 vm.$mount(el)，则代码继续向下执行，如图 3-9 所示。

接着往下看，template 参数选项的有无对生命周期的影响，如例 3-19 所示。

(1) 如果 Vue 实例对象中有 template 参数选项，则将其作为模板编译成 render 函数。

(2) 如果没有 template 选项，则将外部 HTML 作为模板编译。

(3) 可以看到 template 中的模板优先级要高于 outer HTML 的优先级。

【例 3-19】　Vue 对象中增加了 template 选项

```
//第 3 章/Vue 对象中增加了 template 选项.html
<!DOCTYPE html>
```

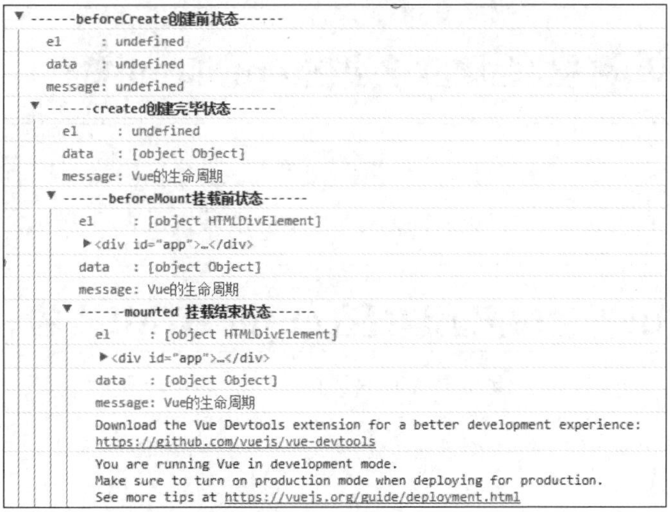

```
▼ ------beforeCreate 创建前状态------
    el      : undefined
    data    : undefined
  message: undefined
▼ ------created 创建完毕状态------
    el      : undefined
    data    : [object Object]
    message: Vue的生命周期
  ▼ ------beforeMount 挂载前状态------
    el      : [object HTMLDivElement]
    ▶ <div id="app">...</div>
    data    : [object Object]
    message: Vue的生命周期
  ▼ ------mounted 挂载结束状态------
    el      : [object HTMLDivElement]
    ▶ <div id="app">...</div>
    data    : [object Object]
    message: Vue的生命周期
Download the Vue Devtools extension for a better development experience:
https://github.com/vuejs/vue-devtools
You are running Vue in development mode.
Make sure to turn on production mode when deploying for production.
See more tips at https://vuejs.org/guide/deployment.html
```

图 3-9　调用 vm.$mount(el)的生命周期

```html
<html lang="en">
<head>
<meta charset="UTF-8">
<meta name="viewport" content="width=device-width, initial-scale=1.0">
<meta http-equiv="X-UA-Compatible" content="ie=edge">
<title>Vue生命周期学习</title>
<script src="https://cdn.bootcss.com/vue/2.4.2/vue.js"></script>
</head>
<body>
<div id="app">
<!-- html 中修改的 -->
<h1>{{message + '这是在 outer HTML 中的'}}</h1>
</div>
</body>
<script>
    var vm = new Vue({
        el: '#app',
        //在 Vue 配置项中修改
            template: "<h1>{{message + '这是在 template 中的'}}</h1>",
            data: {
            message: 'Vue的生命周期'
            }
        })
</script>
</html>
```

运行结果如图 3-10 所示。

Vue的生命周期这是在template中的

图 3-10　在 Vue 对象中增加了 template 选项

将 Vue 对象中 template 的选项注释掉后显示的结果如图 3-11 所示。

Vue的生命周期这是在outer HTML中的

图 3-11　在 Vue 对象中去掉 template 选项

注意：el 进行 DOM 绑定要在 template 之前，因为 Vue 需要通过 el 找到对应的 outer template。

在 Vue 对象中还有一个 render 函数，它是以 createElement 作为参数的，然后进行渲染操作，并且我们可以直接嵌入 JSX，代码如下：

```
var vm = new Vue({
    el: '#app',
    render: function(createElement) {
        return createElement('h1', 'this is createElement')
    }
})
```

运行程序后可以看到页面渲染的结果如图 3-12 所示。

this is createElement

图 3-12　render 函数渲染结果

所以综合排名优先级：render 函数选项 > template 选项 > outer HTML。

（1）beforeMount 和 mounted 钩子函数间的生命周期，如图 3-13 所示。

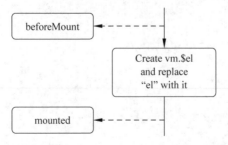

图 3-13　beforeMount 和 mounted 钩子函数间的生命周期

可以看到此时给 Vue 实例对象添加 $el 成员，并且替换掉挂载的 DOM 元素。在之前 console 中打印的结果可以看出 beforeMount 之前 el 的属性还是 undefined。

（2）接下来讲解 mounted 钩子函数的生命周期。

在 mounted 之前 h1 还是通过{{message}}进行占位的，因为此时还没有挂载到页面上，还是在 JavaScript 中以虚拟 DOM 的形式存在。在 mounted 之后可以看到 h1 的内容发生了变化，如图 3-14 所示。

图 3-14　mounted 函数前后内容变化

（3）beforeUpdate 钩子函数和 updated 钩子函数间的生命周期，如图 3-15 所示。

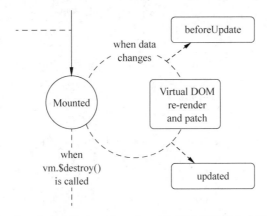

图 3-15　beforeUpdate 和 updated 钩子函数间的生命周期

当 Vue 发现 data 中的数据发生了改变时，会触发对应组件的重新渲染，先后调用 beforeUpdate 和 updated 钩子函数。我们在 console 中输入如下信息：

```
vm.message = '触发组件更新'
```

此时可以发现触发了组件的更新，如图 3-16 所示。

（4）beforeDestroy 和 destroyed 钩子函数间的生命周期，如图 3-17 所示。

beforeDestroy 钩子函数在实例销毁之前调用。在这一步，实例仍然完全可用。而

图 3-16　组件更新状态

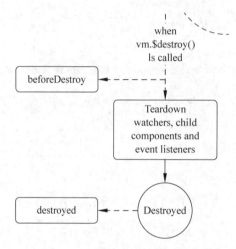

图 3-17　beforeDestroy 和 destroyed 钩子函数间的生命周期

destroyed 钩子函数在 Vue 实例销毁后调用。调用后，Vue 实例指示的所有东西都会被解绑定，所有的事件监听器会被移除，所有的子实例也会被销毁。

3.3　模板渲染

当我们获取后台数据之后，会按照一定的规则加载到前端写好的模板中，并显示在浏览器中，这个过程称为渲染。

3.3.1　条件渲染

1. v-if、v-else-if 和 v-else

v-if、v-else-if、v-else 这 3 个指令后面跟的是表达式。Vue 的条件指令可以根据表达式的值在DOM中渲染或销毁元素/组件,如例3-20所示。

【例3-20】 v-if、v-else-if、v-else 应用

```html
//第3章/条件指令.html
<!DOCTYPE html>
<html xmlns:v-on="http://www.w3.org/1999/xhtml"
xmlns:v-bind="http://www.w3.org/1999/xhtml">
<head>
<title></title>
<meta charset="utf-8"/>
<script src="https://cdn.jsdelivr.net/npm/vue/dist/vue.js"></script>
</head>
<body>
<div id="app">
    <!-- if、else 指令 -->
    <p v-if="status==1">当status为1时,显示该行</p>
    <p v-else-if="status==2">当status为2时,显示该行</p>
    <p v-else-if="status==3">当status为3时,显示该行</p>
    <p v-else>否则显示该行</p>
</div>
<!-- script 脚本 -->
<script>
    //创建Vue实例
    var vm = new Vue({
        el: '#app',
        data: {
            status: 2
        }
    });
</script>
</body>
</html>
```

我们需要注意多个v-if、v-else-if和v-else之间需要紧挨着,如果在它们之间插入一行新代码,则代码如下:

```html
<p v-if="status==1">当status为1时,显示该行</p>
<span></span>
<p v-else-if="status==2">当status为2时,显示该行</p>
<p v-else-if="status==3">当status为3时,显示该行</p>
<p v-else>否则显示该行</p>
```

此时编译浏览器会报错,如图3-18所示。

2. v-show

实际上同v-if效果等同,当绑定事件的元素符合引号中的条件时,该元素才显示,代码如下:

```
 [Vue warn]: Error compiling template:                    vue.js:634
 v-else-if="status==3" used on element <p> without corresponding v-if.
 4 |      <span></span>
 5 |      <p v-else-if="status==2">当status为2时,显示该行</p>
 6 |      <p v-else-if="status==3">当status为3时,显示该行</p>
          ^^^^^^^^^^^^^^^^^^^^^
 7 |      <p v-else="">否则显示该行</p>
 8 |    </div>
 (found in <Root>)
```

图 3-18 v-if、v-else-if 和 v-else 之间不挨着时显示的错误信息

```
<div id="app">
<!-- if、else 指令 -->
<p v-show="status==1">当 status 为 1 时,显示该行</p>
<p v-show="status==2">当 status 为 2 时,显示该行</p>
</div>
<script>
    //创建 Vue 实例
    var vm = new Vue({
        el: '#app',
        data: {
            status: 2
        }
    });
</script>
```

3. v-if 和 v-show 的区别

（1）控制显示的方法不同：该方法和 v-if 区别在于,v-show 实际通过修改 DOM 元素的 display 属性来实现节点的显示和隐藏,而 v-if 则通过添加/删除 DOM 节点来实现。

（2）编译条件：v-if 是惰性的,如果初始条件为假,则什么也不做,此时不会去渲染该元素,只有在条件第一次变为真时才开始局部编译；v-show 则不管初始条件是什么,都被编译,然后被缓存,而且将 DOM 元素保留,只是简单地基于 CSS 进行切换,即 v-if 条件为 true 时才会被加载渲染,而为 false 时不加载。v-show 不管为 true 还是 false,都会被加载渲染。

（3）性能消耗：v-if 有更高的切换消耗,而 v-show 有更高的初始渲染消耗。

（4）使用场景：因此,如果需要非常频繁地切换,则使用 v-show 较好。如果在运行时条件很少改变,只需一次显示或隐藏,则使用 v-if 较好。

4. key

Vue 会尽可能高效地渲染元素,通常会复用已有元素而不是从头开始渲染,如例 3-21 所示,因此可能造成下面这样的问题。

【例 3-21】 Vue 高效地渲染元素

```
//第 3 章/Vue 高效地渲染元素.html
<!DOCTYPE html>
```

```
<html xmlns:v-on="http://www.w3.org/1999/xhtml"
xmlns:v-bind="http://www.w3.org/1999/xhtml">
<head>
<title></title>
<meta charset="utf-8"/>
<script src="https://cdn.jsdelivr.net/npm/vue/dist/vue.js"></script>
</head>
<body>
<div id="app">
    <p v-if="ok">
        <label>Username</label>
        <input placeholder="Enter your username">
    </p>
    <p v-else>
        <label>Email</label>
        <input placeholder="Enter your email address">
    </p>
    <button @click="ok=!ok">切换</button>
</div>
<script type="text/javascript">
    var vm = new Vue({
        el: '#app',
        data: {
            ok: true,
        }
    })
</script>
</body>
</html>
```

页面中输入信息后单击"切换"按钮，此时文本框里的内容并没有清空。

Vue为我们提供了一种方式来声明"这两个元素是完全独立的，不要复用它们"。只需添加一个具有唯一值的key属性，代码如下：

```
<div id="app">
    <p v-if="ok">
        <label>Username</label>
        <input placeholder="Enter your username" key="username-input">
    </p>
    <p v-else>
        <label>Email</label>
        <input placeholder="Enter your email address" key="email-input">
    </p>
    <button @click="ok=!ok">切换</button>
</div>
```

3.3.2 列表渲染

1. v-for 循环用于数组

v-for 指令根据一组数组的选项列表进行渲染。

我们可以用 v-for 指令基于一个数组来渲染一个列表。v-for 指令需要使用 item in items 形式的特殊语法,其中 items 是源数据数组,而 item 则是被迭代的数组元素的别名(为当前遍历的元素提供别名,可以任意起名),如例 3-22 所示。

【例 3-22】 对数组选项进行列表渲染

```html
//第 3 章/数组选项列表渲染.html
<!DOCTYPE html>
<html xmlns:v-on="http://www.w3.org/1999/xhtml"
xmlns:v-bind="http://www.w3.org/1999/xhtml">
<head>
<title></title>
<meta charset="utf-8"/>
<script src="https://cdn.jsdelivr.net/npm/vue/dist/vue.js"></script>
</head>
<body>
<ul id="app">
    <li v-for="item in items">
        {{ item.name }}
    </li>
</ul>
<script type="text/javascript">
    var vm = new Vue({
        el: '#app',
        data: {
            items: [
                { name: 'beixi' },
                { name: 'jzj' }
            ]
        }
    })
</script>
</body>
</html>
```

我们定义一个数组类型的数据 items,用 v-for 将<li>标签循环渲染,效果如图 3-19 所示。

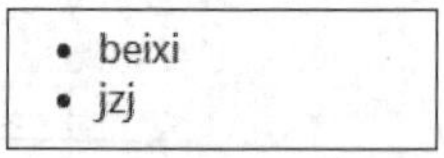

v-for 还支持一个可选的第 2 个参数为当前项的索引,索引从 0 开始,代码如下:

- beixi
- jzj

图 3-19 列表循环结果

```
<ul id = "app">
    <li v-for = "(item,index) in items">
         {{index}} -- {{ item.name }}
    </li>
</ul>
```

分隔符 in 前的语句使用括号，第 2 项就是 items 当前项的索引，运行的结果如图 3-20 所示。

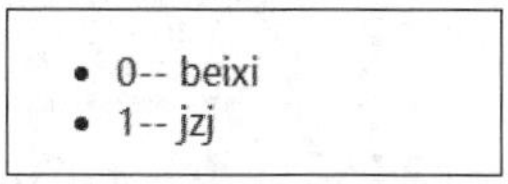

注意：可以用 of 代替 in 作为分隔符。

2. v-for 用于对象

v-for 通过一个对象的属性来遍历并输出结果，如例 3-23 所示。

- 0-- beixi
- 1-- jzj

图 3-20 含有 index 选项的列表渲染结果

【例 3-23】 v-for 来遍历对象

```
//第 3 章/v-for 来遍历对象.html
<!DOCTYPE html>
<html xmlns:v-on = "http://www.w3.org/1999/xhtml"
xmlns:v-bind = "http://www.w3.org/1999/xhtml">
<head>
<title></title>
<meta charset = "utf-8"/>
<script src = "https://cdn.jsdelivr.net/npm/vue/dist/vue.js"></script>
</head>
<body>
<ul id = "app">
    <li v-for = "value in object">
         {{ value }}
    </li>
</ul>
<script type = "text/javascript">
    var vm = new Vue({
        el: '#app',
        data: {
            object: {
                name: 'beixi',
                gender: '男',
                age: 30
            }
        }
    })
</script>
</body>
</html>
```

渲染后的结果如图 3-21 所示。

遍历对象属性时,有两个可选参数,分别是键名和索引,代码如下:

```
<ul id="app">
    <li v-for="(value, key, index) in object">
            {{ index }} -- {{ key }}: {{ value }}
    </li>
</ul>
```

渲染后的结果如图 3-22 所示。

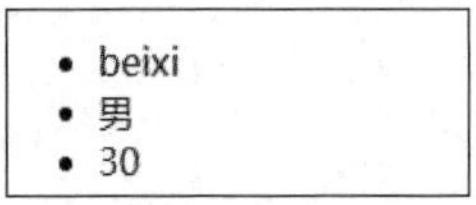

- beixi
- 男
- 30

- 0-- name: beixi
- 1-- gender: 男
- 2-- age: 30

图 3-21 遍历对象结果 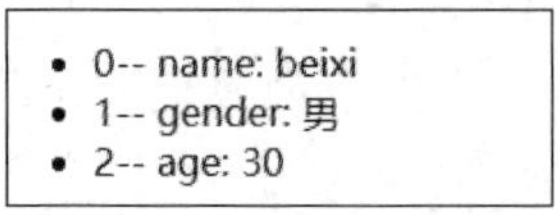　　　　图 3-22 遍历对象的渲染结果

3. v-for 用于整数

v-for 还可以迭代整数,代码如下:

```
<ul id="app">
    <li v-for="n in 10">
            {{n}}
    </li>
</ul>
<script type="text/javascript">
        var vm = new Vue({
            el: '#app',
        })
</script>
```

3.3.3　template 标签用法

上述例子中,v-if 和 v-show 指令都包含在一个根元素中,那是否有方法可以将指令作用到多个兄弟 DOM 元素上呢? Vue 提供了内置标签<template>,可以将多个元素进行渲染,代码如下:

```
<div id="app">
    <div v-if="ok">
        <p>这是第 1 段代码</p>
        <p>这是第 2 段代码</p>
        <p>这是第 3 段代码</p>
    </div>
</div>
```

```
<script type="text/javascript">
    var vm = new Vue({
        el: '#app',
        data:{
            ok:true
        }
    })
</script>
```

同样,<template>标签也支持使用 v-for 指令,用来渲染同级的多个兄弟元素,代码如下:

```
<ul id="app">
<template v-for="item in items">
        {{ item.name }}
        {{ item.age }}
</template>
</ul>
<script type="text/javascript">
    var vm = new Vue({
        el: '#app',
        data: {
            items: [
                { name: 'beixi' },
                { age: 18 }
            ]
        }
    })
</script>
```

3.4 事件绑定

Vue 提供了 v-on 指令来监听 DOM 事件,在事件绑定上,类似原生 JavaScript 的 onclick 事件写法,也是在 HTML 上进行监听。

3.4.1 基本用法

Vue 中的事件绑定,使用 v-on:事件名=函数名来完成,这里函数名定义在 Vue 实例的 methods 对象中,Vue 实例可以直接访问其中的方法。

语法规则:

```
v-on:事件名.修饰符 = 方法名() | 方法名 | 简单的JS表达式
```

> 简写：@事件名.修饰符 = 方法名() | 方法名 | 简单的JS表达式
> 事件名：click|keydown|keyup|mouseover|mouseout|自定义事件名

1. 直接使用

直接在标签中书写js方法，代码如下：

```html
<div id="app">
    单击次数{{count}}
    <button @click="count++">单击+1</button>
</div>
<script type="text/javascript">
    var vm = new Vue({
        el:'#app',
        data:{
            count:0
        }
    })
</script>
```

注意：@click等同于v-on:click，是一个语法糖。

2. 调用methods的方法

通过v-on绑定实例选项属性methods中的方法作为事件的处理器，如例3-24所示。

【例3-24】 v-on绑定事件

```html
//第3章/v-on绑定事件.html
<!DOCTYPE html>
<html xmlns:v-on="http://www.w3.org/1999/xhtml"
    xmlns:v-bind="http://www.w3.org/1999/xhtml">
<head>
<title></title>
<meta charset="utf-8"/>
<script src="https://cdn.jsdelivr.net/npm/vue/dist/vue.js"></script>
</head>
<body>
<div id="app">
    <button @click="say">单击</button>
</div>
<script type="text/javascript">
    var vm = new Vue({
        el:'#app',
        data:{
            msg:'Say Hello'
        },
```

```
        methods:{
            say: function () {
                alert(this.msg)
            }
        }
    })
</script>
</body>
</html>
```

单击 button 按钮,即可触发 say 函数,弹出 alert 框 'Say Hello'。

(1) 方法传参,方法直接在调用时在方法内传入参数,代码如下:

```
<div id = "app">
    <button @click = "say('Hello beixi')">单击</button>
</div>
<script type = "text/javascript">
    var vm = new Vue({
        el: '#app',
        data:{
            msg:'Say Hello'
        },
        methods:{
            say: function (val) {
                alert(val)
            }
        }
    })
</script>
```

(2) 传入事件对象,代码如下:

```
<div id = "app">
<button data - aid = '123' @click = "eventFn( $ event)">事件对象</button>
</div>
<script type = "text/javascript">
    var vm = new Vue({
        el: '#app',
        methods:{
            eventFn:function(e){
                console.log(e);
                // e.srcElement,DOM 节点
                e.srcElement.style.background = 'red';
                console.log(e.srcElement.dataset.aid); /* 获取自定义属性的值 */
```

```
            }
        }
    })
</script>
```

3.4.2 修饰符

Vue 为指令 v-on 提供了多个修饰符,方便我们处理一些 DOM 事件的细节,Vue 主要的修饰符如下。

(1).top:阻止事件继续传播,即阻止它的捕获和冒泡过程。代码如下:

```
@click.stop='handle()'     //只要在事件后面加上.stop 就可以阻止事件冒泡
```

如例 3-25 所示单击"内部单击"按钮,阻止了冒泡过程,即只执行 inner 这个方法,如果不加.stop,则先执行 inner 方法,然后执行 outer 方法,即通过了冒泡这个过程。

【例 3-25】 .stop 修饰符应用

```
//第 3 章/stop 修饰符应用.html
<!DOCTYPE html>
<html xmlns:v-on = "http://www.w3.org/1999/xhtml"
xmlns:v-bind = "http://www.w3.org/1999/xhtml">
<head>
<title></title>
<meta charset = "utf-8"/>
<script src = "https://cdn.jsdelivr.net/npm/vue/dist/vue.js"></script>
</head>
<body>
<div id = "app">
    <div style = "background-color: aqua;width: 100px;height: 100px" v-on:click = "outer">
        外部单击
        <div style = "background-color: red;width: 50px;height: 50px" v-on:click.stop = "inner">内部单击</div>
    </div>
</div>
<script type = "text/javascript">
    var vm = new Vue({
        el: '#app',
        methods:{
            outer: function () {
                console.log("外部单击")
            },
            inner: function () {
```

```
                    console.log("内部单击")
                }
            }
        })
    </script>
</body>
</html>
```

(2).prevent：阻止默认事件。代码如下：

```
@click.prevent = 'handle()'    //只要在事件后面加上.prevent就可以阻止默认事件
```

如下阻止了 a 标签的默认刷新：

```
<a href="" v-on:click.prevent>单击</a>
```

(3).capture：添加事件监听器时使用事件捕获模式，即在捕获模式下触发。代码如下：

```
@click.capture = 'handle()'
```

如下实例在单击最里层的"单击 6"时，outer 方法先执行，因为 outer 方法在捕获模式执行，先于冒泡事件。按下列执行顺序执行：outer→set→inner，因为后两个还是冒泡模式下触发的事件，代码如下：

```
<div v-on:click.capture = "outer">外部单击 5
    <div v-on:click = "inner">内部单击 5
        <div v-on:click = "set">单击 6 </div>
    </div>
</div>
```

(4).self：当前元素自身是触发处理函数时才会触发函数。

原理是根据 event.target 确定是否为当前元素本身，以此来决定是否触发事件/函数。

如下示例，如果单击"内部单击"按钮，冒泡不会执行 outer 方法，因为 event.target 指的是内部单击的 DOM 元素，而不是外部单击的，所以不会触发自己的单击事件，代码如下：

```
<div v-on:click.self = "outer">
    外部单击
    <div v-on:click = "inner">内部单击</div>
</div>
```

(5).once:只触发一次。代码如下:

```
<div id="app">
    <div v-on:click.once="once">单击once</div>
</div>
<script type="text/javascript">
    var vm = new Vue({
        el:'#app',
        methods:{
            once:function(){
                console.log("单击once")
            }
        }
    })
</script>
```

(6)键盘事件。

方式一:@keydown='show($event)'

```
<div id="app">
    <input type="text" @keydown='show($event)'/>
</div>
<script type="text/javascript">
    var vm = new Vue({
        el:'#app',
        methods:{
            show:function(ev){
                /*在函数中获取ev.keyCode*/
                console.log(ev.keyCode);
                if(ev.keyCode==13){
                    alert('你按了回车键!')
                }
            }
        }
    })
</script>
```

方式二:

```
<input type="text" @keyup.enter="show()">          //回车执行
<input type="text" @keydown.up='show()'>           //上键执行
<input type="text" @keydown.down='show()'>         //下键执行
<input type="text" @keydown.left='show()'>         //左键执行
<input type="text" @keydown.right='show()'>        //右键执行
```

3.5 基础 demo 案例

在前面几个章节中,我们介绍了一些 Vue 的基础知识,结合以上知识我们可以做个小 demo:图书管理系统。图书管理系统主要实现数据的添加、删除、列表渲染等功能,最终效果如图 3-23 所示。

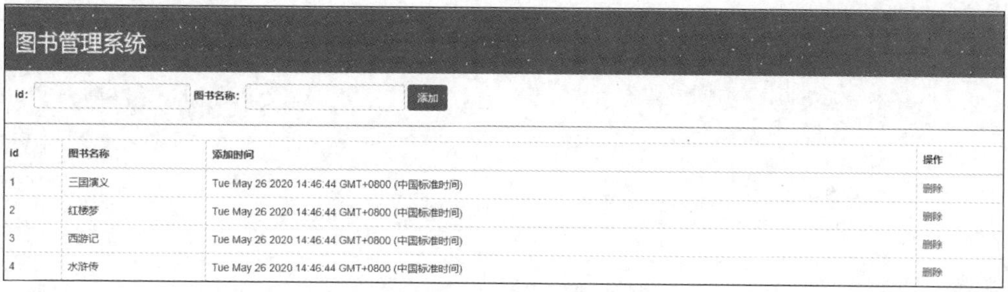

图 3-23 图书管理系统效果图

这个 demo 是基于 BootStrap 快速搭建的,所以对 BootStrap 不是很了解的同学,可以先自行到官网 http://www.bootcss.com/ 进行学习。开始 demo 之前需下载 bootstrap 文件,这里采用的版本是 bootstrap-3.3.7.css。

3.5.1 列表渲染

```
<!DOCTYPE html>
<html xmlns:v-on="http://www.w3.org/1999/xhtml"
xmlns:v-bind="http://www.w3.org/1999/xhtml">
<head>
    <title></title>
    <meta charset="utf-8"/>
    <script src="https://cdn.jsdelivr.net/npm/vue/dist/vue.js"></script>
    <!-- 引入 bootstrap -->
    <link rel="stylesheet" href="./lib/bootstrap-3.3.7.css">
</head>
<body>
<div class="app">
    <div class="panel panel-primary">
        <div class="panel-heading">
            <h2>图书管理系统</h2>
        </div>
        <div class="panel-body form-inline">
            <label for=""> id: <input type="text" class="form-control" v-model="id">
</label>
```

```html
                <label for="">图书名称：<input type="text" class="form-control" v-model="name"></label>
                    <input type="button" value="添加" class="btn btn-primary" @click="add">
            </div>
        </div>
        <table class="table table-bordered table-hover">
            <thead>
                <tr>
                    <th>id</th>
                    <th>图书名称</th>
                    <th>添加时间</th>
                    <th>操作</th>
                </tr>
            </thead>
            <tbody>
                <tr v-for="item in arr" :key="item.id">
                    <td v-text="item.id"></td>
                    <td v-text="item.name"></td>
                    <td v-text="item.time"></td>
                    <td><a href="" @click.prevent="del(item.id)">删除</a></td>
                </tr>
            </tbody>
        </table>
    </div>
<script>
    var vm = new Vue({        //创建Vue实例
        el:'.app',
        data:{
            arr:[
                {'id':1,'name':'三国演义','time':new Date()},
                {'id':2,'name':'红楼梦','time':new Date()},
                {'id':3,'name':'西游记','time':new Date()},
                {'id':4,'name':'水浒传','time':new Date()}
            ],        //创建一些初始数据及其格式
            id:'',
            name:''
        },
    })
</script>
</body>
</html>
```

3.5.2 功能实现

添加功能的代码如下：

```
add(){
    this.arr.push({'id':this.id,'name':this.name,'time':new Date()});
    this.id = this.name = '';
}
```

删除功能的代码如下：

```
del(id){
    var index = this.arr.findIndex(item => {
      if(item.id == id) {
            return true;
        }
    })
    //findIndex方法查找索引,实现列表的删除功能
    this.arr.splice(index,1)
}
```

完整示例代码如下：

```
<!DOCTYPE html>
<html xmlns:v-on="http://www.w3.org/1999/xhtml"
xmlns:v-bind="http://www.w3.org/1999/xhtml">
<head>
<meta charset="utf-8"/>
<script src="https://cdn.jsdelivr.net/npm/vue/dist/vue.js"></script>
<link rel="stylesheet" href="./lib/bootstrap-3.3.7.css">
</head>
<body>
<div class="app">
    <div class="panel panel-primary">
        <div class="panel-heading">
            <h2>图书管理系统</h2>
        </div>
        <div class="panel-body form-inline">
            <label for=""> id: <input type="text" class="form-control" v-model="id"></label>
            <label for="">图书名称: <input type="text" class="form-control" v-model="name"></label>
            <input type="button" value="添加" class="btn btn-primary" @click="add">
        </div>
    </div>
    <table class="table table-bordered table-hover">
        <thead>
        <tr>
            <th>id</th>
```

```html
            <th>图书名称</th>
            <th>添加时间</th>
            <th>操作</th>
        </tr>
    </thead>
    <tbody>
        <tr v-for="item in arr" :key="item.id">
            <td v-text="item.id"></td>
            <td v-text="item.name"></td>
            <td v-text="item.time"></td>
            <td><a href="" @click.prevent="del(item.id)">删除</a></td>
        </tr>
    </tbody>
</table>
</div>
<script>
    var vm = new Vue({        //创建 Vue 实例
        el:'.app',
        data:{
            arr:[
                {'id':1,'name':'三国演义','time':new Date()},
                {'id':2,'name':'红楼梦','time':new Date()},
                {'id':3,'name':'西游记','time':new Date()},
                {'id':4,'name':'水浒传','time':new Date()}
            ],          //创建一些初始数据与格式
            id:'',
            name:''
        },
        methods:{
            add(){
                this.arr.push({'id':this.id,'name':this.name,'time':new Date()});
                this.id = this.name = '';
            },          //add 方法实现列表的输入功能
            del(id){
                var index = this.arr.findIndex(item => {
                    if(item.id == id) {
                        return true;
                    }
                })
/* findIndex 方法查找索引,实现列表的删除功能 */
                this.arr.splice(index,1)
            }
        }
    })
</script>
</body>
</html>
```

第 4 章 自定义指令

除了核心功能默认内置的指令(v-model 和 v-show),Vue 也允许注册自定义指令。注意,在 Vue 2.0 中,代码复用和抽象的主要形式是组件(第 5 章会讲解)。然而,在有的情况下,仍然需要对普通 DOM 元素进行底层操作,这时候就会用到自定义指令。

4.1 指令的注册

自定义指令的注册分为全局注册和局部注册,例如注册一个 v-focus 指令用于<input>、<textarea>元素在页面加载时自动获得焦点。即只要打开这个页面后没单击任何内容,这个页面的输入框就应当处于聚焦状态。

语法:Vue.directive(id,definition)。id 是指令的唯一标识,definition 定义对象则是指令的相关属性及钩子函数,格式如下:

```
// 注册一个全局自定义指令 v-focus
Vue.directive('focus', {
//定义对象
})
```

也可以注册局部指令,组件或 Vue 构造函数中接受一个 directives 的选项,格式如下:

```
var vm = new Vue({
  el:'#app',
  directives:{
     focus:{
//定义对象
     }
  }
})
```

4.2 指令的定义对象

4.1节只是注册了自定义指令v-focus,并没有赋予这个指令任何功能,我们可以传入definition定义对象,从而对指令赋予具体的功能。

一个指令定义对象可以提供以下几个钩子函数(均为可选)。

(1) bind:只调用一次,指令第一次绑定到元素时调用,在这里可以进行一次性初始化设置。

(2) inserted:被绑定元素插入父节点时调用(父节点存在即可调用,不必存在于ocument中)。

(3) update:被绑定元素所在的模板更新时调用,而不论绑定值是否变化。通过比较更新前后绑定的值,可以忽略不必要的模板更新。

(4) componentUpdated:被绑定元素所在模板完成一次更新周期时调用。

(5) unbind:只调用一次,指令与元素解绑时调用。

根据需求在不同的钩子函数内完成逻辑代码,如上面的v-focus,我们希望元素插入父节点时就调用,比较好的钩子函数是inserted,如例4-1所示。

【例4-1】 自定义v-focus指令

```
//第4章/自定义v-focus指令.html
<!DOCTYPE html>
<html xmlns:v-on = "http://www.w3.org/1999/xhtml"
xmlns:v-bind = "http://www.w3.org/1999/xhtml">
<head>
<title></title>
<meta charset = "utf-8"/>
<script src = "https://cdn.jsdelivr.net/npm/vue/dist/vue.js"></script>
</head>
<body>
<div id = "app">
    <p>页面载入时,input元素自动获取焦点:</p>
    <input v-focus >
</div>
<script>
    //注册一个全局自定义指令v-focus
    /* Vue.directive('focus', {
        //当绑定元素插入到DOM中
        inserted: function (el) {
            //聚焦元素
            el.focus()
        }
    }) */
```

```
            // 创建根实例
            var vm = new Vue({
                el: '#app',
                directives: {
                    //注册一个局部的自定义指令 v-focus
                    focus: {
                        //指令的定义
                        inserted: function (el) {
                            //聚焦元素
                            el.focus()
                        }
                    }
                }
            })
        </script>
    </body>
</html>
```

在浏览器中的显示效果如图 4-1 所示。

一旦打开页面，input 输入框就自动获得焦点，成为可输入状态。

图 4-1　v-focus 渲染后的效果

4.3　指令实例属性

除了指令的生命周期外，大家还需知道指令中能调用的相关属性，以便我们对相关 DOM 进行操作。在指令的钩子函数内，可以通过 this 来调用指令实例。下面详细说明指令的实例属性。

（1）el：指令所绑定的元素，可以用来直接操作 DOM。

（2）binding：一个对象，包含以下 property。

- name：指令名，不包括 v-前缀。
- value：指令的绑定值，例如：v-my-directive="1+1"中，绑定值为 2。
- oldValue：指令绑定的前一个值，仅在 update 和 componentUpdated 钩子中可用。无论值是否改变都可用。
- expression：字符串形式的指令表达式。例如 v-my-directive="1+1"中，表达式为 "1+1"。
- arg：传给指令的参数，可选。例如 v-my-directive:foo 中，参数为"foo"。
- modifiers：一个包含修饰符的对象。例如：v-my-directive.foo.bar 中，修饰符对象为{foo: true, bar: true}。

（3）vnode：Vue 编译生成的虚拟节点。移步 VNode API 来了解更多详情。

（4）oldVnode：上一个虚拟节点，仅在 update 和 componentUpdated 钩子中可用。

例 4-2 演示了这些参数的使用。

【例 4-2】 指令实例属性应用

```html
//第 4 章/指令实例属性应用.html
<!DOCTYPE html>
<html xmlns:v-on="http://www.w3.org/1999/xhtml"
xmlns:v-bind="http://www.w3.org/1999/xhtml"
      xmlns:v-demo="http://www.w3.org/1999/xhtml">
<head>
<title></title>
<meta charset="utf-8"/>
<script src="https://cdn.jsdelivr.net/npm/vue/dist/vue.js"></script>
</head>
<body>
<div id="app" v-demo:msg.a.b="message"></div>
</div>
<script>
    Vue.directive('demo', {
        bind: function (el, binding, vnode) {
            var s = JSON.stringify
            el.innerHTML =
                'name: ' + s(binding.name) + '<br>' +
                'value: ' + s(binding.value) + '<br>' +
                'expression: ' + s(binding.expression) + '<br>' +
                'argument: ' + s(binding.arg) + '<br>' +
                'modifiers: ' + s(binding.modifiers) + '<br>' +
                'vnode keys: ' + Object.keys(vnode).join(', ')
        }
    })
    new Vue({
        el: '#app',
        data: {
            message: 'hello!'
        }
    })
</script>
</body>
</html>
```

页面显示结果如图 4-2 所示。

从输出的结果可以看出，在自定义指令中能收到传入的参数和元素等。

有时候我们不需要其他钩子函数，此时可以简写函数，指令函数可接受所有合法的 JavaScript 表达式，以下实例传入了 JavaScript 对象，代码如下：

```
name: "demo"
value: "hello!"
expression: "message"
argument: "msg"
modifiers: {"a":true,"b":true}
vnode keys: tag, data, children, text, elm, ns, context, fnContext, fnOptions, fnScopeId, key,
componentOptions, componentInstance, parent, raw, isStatic, isRootInsert, isComment, isCloned,
isOnce, asyncFactory, asyncMeta, isAsyncPlaceholder
```

图 4-2　指令实例属性执行结果

```
<div id = "app">
    <div v-demo = "{ color: 'green', text: '自定义指令!' }"></div>
</div>
<script>
    Vue.directive('demo', function (el, binding) {
        //以简写方式设置文本及背景颜色
        el.innerHTML = binding.value.text
        el.style.backgroundColor = binding.value.color
    })
    new Vue({
        el: '#app'
    })
</script>
```

Vue 2.0 之后移除了大量 Vue 1.0 自定义指令的配置。在使用自定义指令时，应该充分理解业务需求，因为很多时候需要的可能并不是自定义指令，而是组件。在第 5 章我们将详细学习组件。

4.4　案例

4.4.1　下拉菜单

网页中有很多常见的下拉菜单，当单击"下拉"按钮时，会弹出一个下拉菜单，如例 4-3 所示。

【例 4-3】　下拉菜单

```
//第 4 章/下拉菜单.html
<!DOCTYPE html>
<html>
<head>
<title></title>
<meta charset = "utf-8"/>
```

```html
<script src="https://cdn.jsdelivr.net/npm/vue/dist/vue.js"></script>
</head>
<body>
<div id="app">
    <!-- 自定义指令 v-clickoutside 绑定 handleHide 函数 -->
    <div class="main" v-clickoutside="handleHide">
    <button @click="show = !show">单击显示下拉菜单</button>
        <div class="dropdown" v-show="show">
            <div class="item"><a href="#">选项 1</a></div>
            <div class="item"><a href="#">选项 2</a></div>
            <div class="item"><a href="#">选项 3</a></div>
        </div>
    </div>
</div>
<script>
        /*自定义指令 v-clickoutside*/
        Vue.directive('clickoutside', {
            /*在 document 上绑定 click 事件,所以在 bing 钩子函数内声明了一个函数
            documentHandler,并将它作为句柄绑定在 document 的 click 事件上
            documentHandler 函数做了两个判断。
            **/
            bind(el, binding) {
                function documentHandler(e) {
                    /*第1个是判断单击的区域是否是指令所在的元素内部。如果是,
                    就跳转函数,不往下继续执行。
                    * contains 方法用来判断元素 A 是否包含了元素 B,包含则返回 true
                    **/
                    if (el.contains(e.target)) {
                        return false
                    }
                    /*第2个是判断当前的指令 v-clickoutside 有没有表达式,在该自定义
                    指令中,表达式应该是一个函数,在过滤了内部元素后,单击外面任何区
                    域都应该执行用户表达式中的函数,所以 binding.value()
                    * 用来执行当前上下文 methods 中指定的函数
                    **/
                    if (binding.expression) {
                        binding.value(e)
                    }
                }
                /*与 Vue 1.0 不同的是,在自定义指令中,不能再使用 this.xxx 的形式在上下
                文中声明一个变量,所以用了 el.__vueMenuHandler__ 引用 documentHandler,
                这样就可以在 unbind 钩子里移除对 document 的 click
                事件监听。如果不移除,当组建或元素销毁时,它仍然存在于内存中
                **/
                el.__vueMenuHandler__ = documentHandler;
```

```
                document.addEventListener('click', el.__vueMenuHandler__)
            },
            unbind(el) {
                document.removeEventListener('click', el.__vueMenuHandler__)
                delete el.__vueMenuHandler__
            }
        })
        new Vue({
            el: '#app',
            data: {
                show: false
            },
            methods: {
                handleHide() {
                    this.show = false
                }
            }
        })
    </script>
</body>
</html>
```

页面最终呈现的效果如图 4-3 所示。

示例中逻辑很简单,单击按钮时显示 class 为 dropdown 的 div 元素。自定义指令 v-clickoutside 绑定 handleHide 函数,用来关闭菜单。

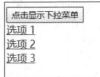

图 4-3　下拉菜单示例最终效果

4.4.2　相对时间转换

在很多社区网站,例如朋友圈、微博等,发布的动态有一个相对本机时间转换后的相对时间,如图 4-4 中圈出的时间。

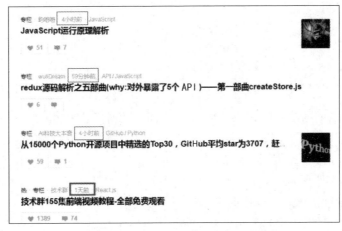

图 4-4　相对本机时间转换后的相对时间

一般在服务器的存储时间格式是 UNIX 时间戳,例如 2018-01-17 06:00:00 的时间戳是 1516140000。前端在得到数据后,将它转换为可持续的时间格式再显示出来。为了显示出实时性,在一些社交类产品中,甚至会实时转换为几秒前、几分钟前、几小时前等不同的格式,因为这样比直接转换为年、月、日、时、分、秒显得对用户更加友好,体验更人性化。

我们来实现这样一个 Vue 自定义指令 v-time,将表达式传入的时间戳实时转换为相对时间。为了便于验证演示效果,我们初始化时定义了两个时间。

index.html 页面代码如下:

```html
<!DOCTYPE html>
<html>
<head>
<title>时间转换指令</title>
<meta charset="utf-8"/>
<script src="https://cdn.jsdelivr.net/npm/vue/dist/vue.js"></script>
</head>
<body>
<div id="app">
    <div v-time="timeNow"></div>
    <div v-time="timeBefore"></div>
</div>
<script src="./time.js"></script>
<script src="./index.js"></script>
</body>
</html>
```

index.js 代码如下:

```js
var vm = new Vue({
    el: '#app',
    data: {
        timeNow: (new Date()).getTime(),
        timeBefore: 1580503571085
    }
})
```

timeNow 是当前的时间,timeBefore 是一个固定的时间:2020-02-01。

分析一下时间转换的逻辑:

(1) 1 分钟以前,显示"刚刚"。

(2) 1 分钟~1 小时,显示"xx 分钟前"。

(3) 1 小时~1 天,显示"xx 小时前"。

(4) 1 天~1 个月(31 天),显示"xx 天前"。

(5) 大于 1 个月,显示"xx 年 xx 月 xx 日"。

这样罗列出来，逻辑就一目了然了。为了使判断更简单，我们这里统一使用时间戳进行大小判断。在写指令 v-time 之前，需要先写一系列与时间相关的函数，我们声明一个对象 Time，把它们都封装到里面。

time.js 代码如下：

```javascript
var Time = {
    //获取当前时间戳
    getUNIX: function() {
        var date = new Date();
        return date.getTime();
    },
    //获取今天 0 点 0 分 0 秒的时间戳
    getTodayUNIX: function() {
        var date = new Date();
        date.setHours(0);
        date.setMinutes(0);
        date.setSeconds(0);
        date.setMilliseconds(0);
        return date.getTime();
    },
    //获取今年 1 月 1 日 0 点 0 秒的时间戳
    getYearUNIX: function() {
        var date = new Date();
        date.setMonth(0);
        date.setDate(1);
        date.setHours(0);
        date.setMinutes(0);
        date.setSeconds(0);
        date.setMilliseconds(0);
        return date.getTime();
    },
    //获取标准年、月、日
    getLastDate: function(time) {
        var date = new Date(time);
        var month = date.getMonth() + 1 < 10 ? '0' + (date.getMonth() + 1) : date.getMonth() + 1;
        var day = date.getDate() < 10 ? '0' + date.getDate() : date.getDate();
        return date.getFullYear() + '-' + month + '-' + day;
    },
    //转换时间
    getFormateTime: function(timestamp) {
        var now = this.getUNIX();
        var today = this.getTodayUNIX();
        var year = this.getYearUNIX();
        var timer = (now - timestamp) / 1000;
```

```
        var tip = '';
        if (timer <= 0) {
            tip = '刚刚';
        } else if (Math.floor(timer / 60) <= 0) {
            tip = '刚刚';
        } else if (timer < 3600) {
            tip = Math.floor(timer / 60) + '分钟前';
        } else if (timer >= 3600 && (timestamp - today >= 0)) {
            tip = Math.floor(timer / 3600) + '小时前';
        } else if (timer / 86400 <= 31) {
            tip = Math.ceil(timer / 86400) + '天前';
        } else {
            tip = this.getLastDate(timestamp);
        }
        return tip;
    }
}
```

Time.getFormatTime()方法就是自定义指令 v-time 所需要的方法,参数为毫秒级时间戳,返回已经整理好的时间格式的字符串。

最后在 time.js 里补全剩余的代码如下:

```
Vue.directive('time', {
    bind: function(el,binding) {
        el.innerHTML = Time.getFormateTime(binding.value);
        el.__timeout__ = setInterval(() => {
            el.innerHTML = Time.getFormateTime(binding.value);
        },60000);
    },
    unbind: function() {
        clearInterval(el.__timeout__);
        delete el.__timeout__;
    }
})
```

在 bind 钩子里,将指令 v-time 表达式的值 binding.value 作为参数传入 Time.getFormatTime()方法中得到格式化时间,再通过 el.innerHTML 写入指令所在元素。定时器 el.__timeout__ 每分钟触发一次,更新时间,并且在 unbind 钩子里清除掉。

总结:在编写自定义指令时,给 DOM 绑定一次性事件等初始动作,建议在 bind 钩子内完成,同时要在 unbind 内解除相关绑定。

第 5 章 组 件

组件(Component)是 Vue 最核心的功能部分,也是整个框架设计最精彩的地方,当然也是比较难掌握的。每个开发者都想在软件开发过程中使用之前写好的代码,但又担心引入这段代码会对现有的程序产生影响。Web Components 的出现提供了一种新的思路,可以自定义 tag 标签,并拥有自身的模板、样式和交互。

5.1 什么是组件

在正式介绍组件前,先看一个 Vue 组件的简单示例感受一下,代码如下:

```
// 定义一个名为 button-counter 的新组件
Vue.component('button-counter', {
  data: function () {
    return {
      count: 0
    }
  },
  template: '<button v-on:click = "count++"> You clicked me {{ count }} times.</button>'
})
```

组件是可复用的 Vue 实例,且带有一个名字,在这个示例中组件名字是< button-counter >。可以在一个通过 new Vue 创建的 Vue 根实例中,把这个组件作为自定义标签来使用,代码如下:

```
< div id = "components-demo">
< button-counter ></button-counter >
< button-counter ></button-counter >
</div >
new Vue({ el: '#components-demo' })
```

完整示例代码如下:

```html
//第5章/自定义组件.html
<!DOCTYPE html>
<html>
<head>
<title></title>
<meta charset="utf-8"/>
<script src="https://cdn.jsdelivr.net/npm/vue/dist/vue.js"></script>
</head>
<body>
<div id="components-demo">
    <button-counter></button-counter>
    <button-counter></button-counter>
</div>
<script>
    //定义一个名为 button-counter 的新组件
    Vue.component('button-counter', {
        data: function () {
            return {
                count: 0
            }
        },
        template: '<button v-on:click="count++">You clicked me {{ count }} times.
          </button>'
    })
    new Vue({ el: '#components-demo' })
</script>
</body>
</html>
```

这些类似于<button-counter>，平时工作中没见过的标签就是组件，每个标签代表一个组件，这样就可以将组件进行任意次数的复用。

Web 的组件其实就是页面组成的一部分，好比是计算机中的每一个硬件（如硬盘、键盘、鼠标等），它具有独立的逻辑和功能或界面，同时又能根据规定的接口规则进行相互融合，从而变成一个完整的应用。

Web 页面就是由一个个类似这样的部分组成的，例如导航、列表、弹窗、下拉菜单等。页面只不过是这些组件的容器，组件自由组合形成功能完整的界面，当不需要某个组件或者想要替换某个组件时，可以随时进行替换和删除，而不影响整个应用的运行。

前端组件化的核心思路就是将一个巨大且复杂的页面分成粒度合理的页面组成部分。

使用组件的好处：

(1) 提高开发效率。

(2) 方便重复使用。

(3) 简化调试步骤。

（4）提升整个项目的可维护性。

（5）便于协同开发。

组件是 Vue.js 最强大的功能之一。组件可以扩展 HTML 元素，封装可重用的代码。在较高层面上，组件是自定义元素，Vue.js 的编译器为它添加特殊功能。在有些情况下，组件也可以以原生 HTML 元素的形式存在，以 is 特性扩展。

组件系统让我们可以用独立可复用的小组件构建大型应用，几乎任意类型应用的界面可以抽象为一个组件树，如图 5-1 所示。

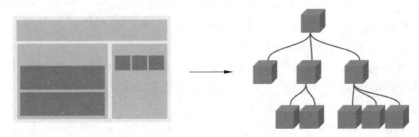

图 5-1　Vue 组件树

5.2　组件的基本使用

为了能在模板中使用，这些组件必须先注册以便 Vue 能够识别。这里有两种组件的注册类型：全局注册和局部注册。

5.2.1　全局注册

全局注册需要确保在根实例初始化之前注册，这样才能使组件在任意实例中被使用。全局注册有 3 种方式。

要注册一个全局组件，我们可以使用 Vue.component(tagName, options)，代码如下：

```
Vue.component('my-component', {
    //选项
})
```

my-component 就是注册组件的自定义标签名称，推荐使用小写字母加 "-" 分隔符的形式命名。组件在注册之后，便可以在父实例的模块中以自定义元素的形式使用，示例代码如下：

```
<div id="app">
    <my-component></my-component>
</div>
<script>
```

```
    Vue.component('my-component', {
        template: '<h1>注册</h1>'
    });
    var vm = new Vue({
        el:'#app'
    })
</script>
```

注意:

(1) template 的 DOM 结构必须被一个而且是唯一根元素所包含,直接引用而不被"<div></div>"包裹是无法被渲染的。

(2) 模板(template)声明了数据和最终展现给用户的 DOM 之间的映射关系。

除了 template 选项外,组件中还可以像 Vue 实例那样使用其他的选项(5.3 节将详细讲解),例如 data、computed、methods 等,但是在使用 data 时和实例不同,data 必须是函数,然后将数据利用 return 返回,代码如下:

```
<div id="app">
    <my-component></my-component>
</div>
<script>
    Vue.component('my-component', {
        template: '<h1>{{message}}</h1>',
        data:function () {
            return{
                message:'注册'
            }
        }
    });
    var vm = new Vue({
        el:'#app'
    })
</script>
```

Vue 组件中 data 值不能为对象,因为对象是引用类型,组件可能会被多个实例同时引用。如果 data 值为对象,将导致多个实例共享一个对象,其中一个组件改变 data 属性值,其他实例也会受影响。

上面解释了 data 不能为对象的原因,这里简单地讲一下 data 为函数的原因。data 为函数,通过 return 返回对象的复制,致使每个实例都有自己独立的对象,实例之间可以互不影响地改变 data 属性值。

使用 Vue.extend 配合 Vue.component 方法,代码如下:

```
<div id="app">
    <my-list></my-list>
</div>
<script>
    var list = Vue.extend({
        template:'<h1>this is a list</h1>',
    });
    Vue.component("my-list",list);
    //根实例
    new Vue({
            el:"#app",
        })
</script>
```

将模板字符串定义到 script 标签中,代码如下:

```
<script id="tmpl" type="text/x-template">
    <div><a href="#">登录</a> | <a href="#">注册</a></div>
</script>
```

同时,需要使用 Vue.component 定义组件,代码如下:

```
Vue.component('account', {
    template: '#tmpl'
});
```

完整示例代码如下:

```
//第5章/全局注册.html
<div id="app">
    <account></account>
</div>
<template id="tmpl">
    <div><a href="#">登录</a> | <a href="#">注册</a></div>
</template>
<script>
    Vue.component('account', {
        template: '#tmpl'
    });
    new Vue({
        el:"#app",
    })
</script>
```

5.2.2 局部注册

如果不需要全局注册或者只让组件使用在其他组件内，可以采用选项对象的components属性实现局部注册，示例代码如下：

```html
//第5章/局部注册.html
<div id="app">
    <account></account>
</div>
<script>
    //创建Vue实例，得到ViewModel
    var vm = new Vue({
        el: '#app',
        data: {},
        methods: {},
        components: {          //定义子组件
            account: {          //account组件
                template: '<div><h1>这是Account组
                </h1><login></login></div>', //在这里使用定义的子组件
                components: {   //定义子组件的子组件
                    login: {    //login组件
                        template: "<h3>这是登录组件</h3>"
                    }
                }
            }
        }
    });
</script>
```

可以使用flag标识符结合v-if和v-else切换组件，如例5-1所示。

【例5-1】 使用标识符切换组件

```html
//第5章/使用标识符切换组件.html
<!DOCTYPE html>
<html>
<head>
<title></title>
<meta charset="utf-8"/>
<script src="https://cdn.jsdelivr.net/npm/vue/dist/vue.js"></script>
</head>
<body>
<div id="app">
    <input type="button" value="toggle" @click="flag=!flag">
    <account v-if="flag"></account>
```

```
        <login v-else="flag"></login>
    </div>
    <script>
        //创建 Vue 实例,得到 ViewModel
        var vm = new Vue({
            el: '#app',
            data: {
                flag: true
            },
            methods: {},
            components: {        //定义子组件
                account: {       //account 组件
                    template: '<div><h1>这是 Account 组件</h1></div>',
                                                      // 在这里使用定义的子组件
                },
                login: {         //login 组件
                    template: "<h3>这是登录组件</h3>"
                }
            }
        });
    </script>
</body>
</html>
```

5.2.3 DOM 模板解析说明

当使用 DOM 作为模版时(例如,将 el 选项挂载到一个已存在的元素上),会受到 HTML 的一些限制,因为 Vue 只有在浏览器解析和标准化 HTML 后才能获取模板内容。尤其像这些元素 <ul>、<ol>、<table>、<select> 限制了能被它包裹的元素,而一些像 <option>这样的元素只能出现在某些其他元素内部。

在自定义组件中使用这些受限制的元素时会导致一些问题,例如:

```
<table>
    <my-row>...</my-row>
</table>
```

自定义组件有时被认为是无效的内容,因此在渲染的时候会导致错误。这时要使用特殊的 is 属性来挂载组件,代码如下:

```
<table>
    <tr is="my-row"></tr>
</table>
```

也就是说,在标准 HTML 中,一些元素只能放置特定的子元素,而另一些元素只能存在于特定的父元素中。例如 table 中不能放置 div,tr 的父元素不能为 div 等。所以,当使用自定义标签时,标签名还是那些标签的名字,但是可以在标签的 is 属性中填写自定义组件的名字,如例 5-2 所示。

【例 5-2】 DOM 模板解析示例

```
//第 5 章/DOM 模板解析.html
<!DOCTYPE html>
<html>
<head>
<title></title>
<meta charset="utf-8"/>
<script src="https://cdn.jsdelivr.net/npm/vue/dist/vue.js"></script>
</head>
<body>
<div id="app">
    <table border="1" cellpadding="5" cellspacing="0">
        <my-row></my-row>
        <tr is="my-row"></tr>
    </table>
</div>
<script type="text/javascript">
    new Vue({
        el:'#app',
        components:{
            myRow:{
                template:'<tr><td>123456</td></tr>'
            }
        }
    });
</script>
</body>
</html>
```

示例执行效果如图 5-2 所示。

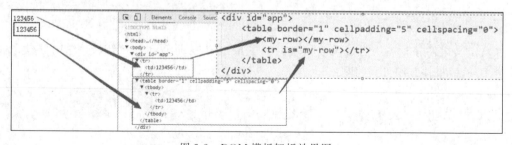

图 5-2 DOM 模板解析效果图

从图中不难发现直接引用< my-row >组件标签并没有被< table >标签包裹，而用 is 特殊属性挂载的组件可以达到所需效果。

注意：如果使用的是字符串模板，则不受限制。

5.3 组件选项

Vue 的组件可以理解为预先定义好行为的 ViewModel 类。一个组件可以预先定义很多选项。但是最核心的有以下几个。

（1）模板(template)：模板声明了数据和最终展现给用户的 DOM 之间的映射关系。

（2）初始数据(data)：一个组件的初始数据状态。对于可复用的组件来说，通常是私有的状态。

（3）接收的外部参数(props)：组件之间通过参数来进行数据的传递和共享。参数默认是单向绑定（由上至下），但也可以显式声明为双向绑定。

（4）方法(methods)：对数据的改动操作一般都在组件的方法内进行。可以通过 v-on 指令将用户输入事件和组件方法进行绑定。

（5）生命周期钩子函数(lifecycle hooks)：一个组件会触发多个生命周期钩子函数，例如 created、attached、destroyed 等。在这些钩子函数中，我们可以封装一些自定义的逻辑。和传统的 MVC 相比，这可以理解为 Controller 的逻辑被分散到了这些钩子函数中。

组件接收的选项大部分与 Vue 实例一样，相同的部分就不再赘述了。我们重点讲解一下二者不同的选项 data 和 props，data 在 5.2.1 节中已经讲解过了，所以本节主要讲解 props，它用于接收父组件传递的参数。

5.3.1 组件 props

上述我们在使用组件时主要是把组件模板的内容进行复用，代码如下：

```
//父组件
<div id="app">
    <my-component></my-component>
</div>
<script>
    //子组件
    Vue.component('my-component', {
        template: '<h1>注册</h1>'
    });
    var vm = new Vue({
        el:'#app'
    })
</script>
```

但是组件中更重要的功能是组件间进行通信，选项 props 是组件中非常重要的一个选项，起到父子组件间桥梁的作用。

1. 静态 props

组件实例的作用域是孤立的。这意味着不能（也不应该）在子组件的模板内直接引用父组件的数据，代码如下：

```
<div id="app">
    <my-componet message="来自父组件的数据!"></my-componet>
</div>
<script type="text/javascript">
    Vue.component('my-componet', {
        template: '<span>{{ message }}</span>'
    })
    new Vue({
        el:'#app'
    });
</script>
```

子组件接收不到父组件 message 的数据，而且会报错，如图 5-3 所示。

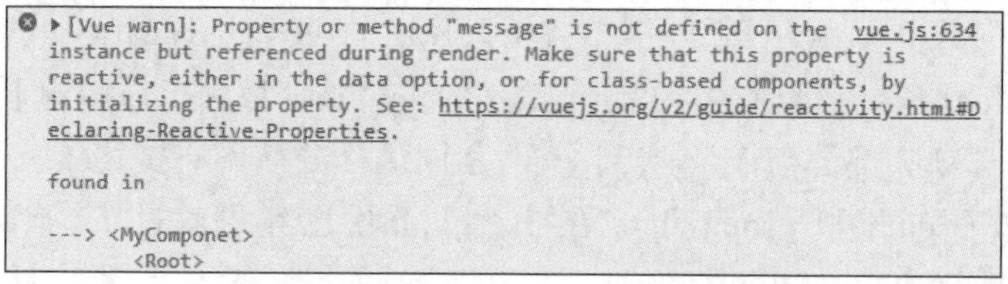

```
⊗ ▶[Vue warn]: Property or method "message" is not defined on the    vue.js:634
   instance but referenced during render. Make sure that this property is
   reactive, either in the data option, or for class-based components, by
   initializing the property. See: https://vuejs.org/v2/guide/reactivity.html#D
   eclaring-Reactive-Properties.

   found in

   ---> <MyComponet>
          <Root>
```

图 5-3　直接引用父组件 message 后的错误信息

要想让子组件使用父组件的数据，需要通过子组件的 props 选项实现。子组件要显式地用 props 选项声明它期望获得的数据，如例 5-3 所示。

【例 5-3】　子组件使用父组件的数据

```
//第5章/子组件使用父组件的数据.html
//部分代码省略
<div id="app">
    <my-componet message="来自父组件的数据!"></my-componet>
</div>
<script type="text/javascript">
    Vue.component('my-componet', {
        //声明 props
        props: ['message'],
        //就像 data 一样,props 可以用在模板内
        //同样也可以在 vm 实例中像 "this.message" 这样使用
```

```
        template:'<span>{{ message }}</span>'
    })
    new Vue({
        el:'#app'
    });
</script>
```

页面显示结果为：来自父组件的数据！

由于 HTML 的特性不区分大小写，所以，当使用的不是字符串模板时，camelCased（驼峰式）命名的 props 需要转换为相对应的 kebab-case（短横线隔开式）命名，代码如下：

```
<div id="app">
    <my-componet my-message="来自父组件的数据！"></my-componet>
</div>
<script type="text/javascript">
    Vue.component('my-componet',{
        props:['myMessage'],
        template:'<span>{{ myMessage }}</span>'
    })
    new Vue({
        el:'#app'
    });
</script>
```

2. 动态 props

在模板中，有时候传递的数据并不一定是固定不变的，而是要动态地绑定父组件的数据到子模板的 props，与绑定到任何普通的 HTML 特性相类似，采用 v-bind 进行绑定。当父组件的数据变化时，该变化也会传递给子组件，如例 5-4 所示。

【例 5-4】 动态传递父组件数据到子组件

```
//第 5 章/动态 props.html
<!DOCTYPE html>
<html>
<head>
<title></title>
<meta charset="utf-8"/>
<script src="https://cdn.jsdelivr.net/npm/vue/dist/vue.js"></script>
</head>
<body>
<div id="app">
    <input type="text" v-model="parentMessage">
    <my-componet :message="parentMessage"></my-componet>
</div>
```

```
<script type="text/javascript">
    Vue.component('my-componet', {
        props: ['message'],
        template: '<span>{{ message }}</span>'
    })
    new Vue({
        el:'#app',
        data:{
            parentMessage:''
        }
    });
</script>
</body>
</html>
```

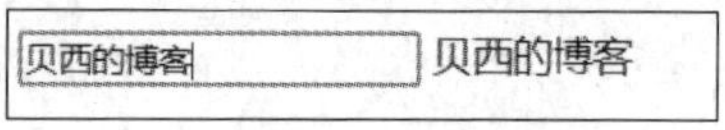

图 5-4 动态接收父组件数据效果图

页面显示效果如图 5-4 所示。

这里使用 v-model 绑定了父组件数据 parentMessage，当在输入框中输入数据时，子组件接收到的 props 'message' 也会实时响应，并更新组件模板。

对于初学者常犯的一个错误来讲，如果在父组件中直接传递数字、布尔值、数组、对象，则它所传递的值均为字符串。如果想传递一个实际的值，则需要使用 v-bind，从而让它的值被当作 JavaScript 表达式进行计算，代码如下：

```
<div id="app">
    <my-componet message="1+1"></my-componet><br>
    <my-componet :message="1+1"></my-componet>
</div>
<script type="text/javascript">
    Vue.component('my-componet', {
        props: ['message'],
        template: '<span>{{ message }}</span>'
    })
    new Vue({
        el:'#app'
    });
</script>
```

如果不使用 v-bind 指令，则页面显示结果是字符串"1+1"。如果使用 v-bind 指令，则会当作 JavaScript 表达式进行计算，其计算结果是数字 2。

5.3.2 props 验证

前面我们介绍的 props 选项的值都是数组，除了数组，还可以是对象。可以为组件的

props 指定验证规则，如果传入的数据不符合规则，则 Vue 会发出警告。当 props 需要验证时，需要采用对象写法。

当组件给其他人使用时，推荐进行数据验证。以下是一些 props 的示例代码：

```
Vue.component('example', {
    props: {
        //基础类型检测,null 的意思是任何类型都可以
        propA: Number,
        //多种类型
        propB: [String, Number],
        //必传且是字符串
        propC: {
            type: String,
            required: true
        },
        //数字,有默认值
        propD: {
            type: Number,
            default: 100
        },
        //数组/对象的默认值应当由一个工厂函数返回
        propE: {
            type: Object,
            default: function () {
                return { message: 'hello' }
            }
        },
        //自定义验证函数
        propF: {
            validator: function (value) {
                return value > 10
            }
        }
    }
})
```

验证的 type 类型可以是 String、Number、Boolean、Function、Object、Array 等，type 也可以是一个自定义构造器函数，使用 instanceof 检测。

当 props 验证失败时，Vue 会抛出警告（如果使用的是开发版本）。props 会在组件实例创建之前进行校验，所以在 default 或 validator 函数里，诸如 data、computed 或 methods 等实例属性还无法使用。

【例 5-5】 如果传入子组件的 message 不是数字，则抛出警告

```
//第 5 章/props 验证.html
```

```html
<!DOCTYPE html>
<html>
<head>
<title></title>
<meta charset="utf-8"/>
<script src="https://cdn.jsdelivr.net/npm/vue/dist/vue.js"></script>
</head>
<body>
<div id="example">
    <parent></parent>
</div>
<script>
    var childNode = {
        template: '<div>{{message}}</div>',
        props:{
            'message':Number
        }
    }
    var parentNode = {
        template:'<div class="parent"><child :message="msg"></child></div>',
        components: {
            'child': childNode
        },
        data(){
        return{
            msg: '123'
        }
      }
    };
    new Vue({      //创建根实例
        el: '#example',
        components: {
            'parent': parentNode
        }
    })
</script>
</body>
</html>
```

传入数字 123 时,无警告提示。但当传入字符串"123"时,控制台会发出警告,如图 5-5 所示。

5.3.3 单向数据流

所有的 props 都使得其父子 props 之间形成了一个单向下行绑定:父级 props 的更新

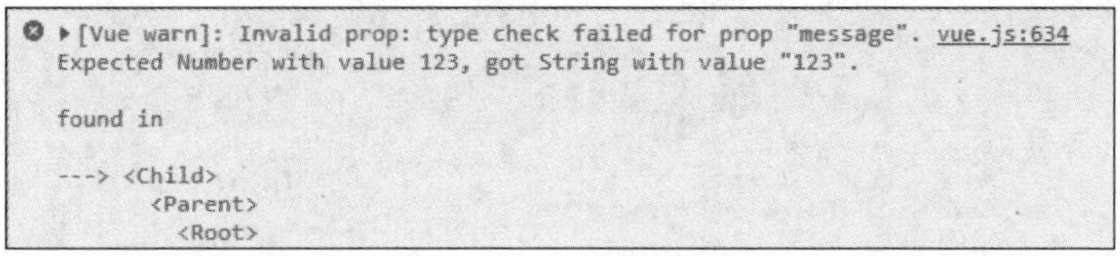

```
[Vue warn]: Invalid prop: type check failed for prop "message".    vue.js:634
Expected Number with value 123, got String with value "123".

found in

---> <Child>
       <Parent>
         <Root>
```

图 5-5　props 验证失败时的警告信息

会向下流动到子组件中,但是反过来则不行。之所以这样设计,是尽可能将父子组件解耦,避免子组件无意中修改了父组件的状态。

额外地,每次父级组件发生变更时,子组件中所有的 props 都将刷新为最新的值。这意味着不应该在一个子组件内部改变 props。如果这样做了,Vue 会在浏览器的控制台中发出警告,如例 5-6 所示。

【例 5-6】　单向数据传递

```
//第5章/单向数据传递.html
<!DOCTYPE html>
<html>
<head>
<title></title>
<meta charset = "utf-8"/>
<script src = "https://cdn.jsdelivr.net/npm/vue/dist/vue.js"></script>
</head>
<body>
<div id = "example">
    <parent></parent>
</div>
<script>
    var childNode = {
        template:
        '<div class = "child"><div><span>子组件数据</span>' +
            '<inputv-model = "childMsg"></div><p>{{childMsg}}</p></div>',
        props:['childMsg']
    }
    var parentNode = {
        template:
        '<div class = "parent"><div><span>父组件数据</span>' +
            '<inputv-model = "msg"></div><p>{{msg}}</p><child :child-msg = "msg">
        </child></div>',
        components: {
            'child': childNode
        },
        data(){
            return {
```

```
                'msg':'match'
            }
        }
    };
    new Vue({         //创建根实例
        el: '#example',
        components: {
            'parent': parentNode
        }
    })
</script>
</body>
</html>
```

页面显示效果如图 5-6 所示。

父组件数据变化时,子组件数据会相应变化,而子组件数据变化时,父组件数据保持不变,并在控制台显示警告,如图 5-7 所示。

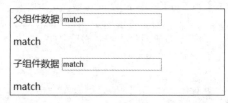

图 5-6　单向数据传递页面效果

业务中经常会遇到需要修改 props 中的数据的情况,通常有以下两种原因:

(1) props 作为初始值传入后,子组件想把它当作局部数据来使用。

(2) props 作为初始值传入,由子组件处理成其他数据并输出。

```
⊗ ▶[Vue warn]: Avoid mutating a prop directly since the value will   vue.js:634
    be overwritten whenever the parent component re-renders. Instead, use a data
    or computed property based on the prop's value. Prop being mutated:
    "childMsg"
```

图 5-7　修改子组件时的警告信息

注意:JS 中对象和数组是引用类型,指向同一个内存空间,如果 props 是一个对象或数组,在子组件内部改变它会影响父组件的状态。

对于这两种情况,正确的应对方式是:

(1) 定义一个局部变量,并用 props 的值初始化它,代码如下:

```
props: ['initialCounter'],
data: function () {
  return { counter: this.initialCounter }
}
```

但是,定义的局部变量 counter 只能接收 initialCounter 的初始值,当父组件要传递的值发生变化时,counter 无法接收到最新值,示例代码如下:

```html
<!DOCTYPE html>
<html lang="en">
<head>
<meta charset="UTF-8">
<title>Title</title>
</head>
<body>
<div id="example">
    <parent></parent>
</div>
<script>
    var childNode = {
        template:
        '<div class="child"><div><span>子组件数据</span>' +
            '<input v-model="temp"></div><p>{{temp}}</p></div>',
        props:['childMsg'],
        data(){
            return{
                temp:this.childMsg
            }
        },
    };
    var parentNode = {
        template:
        '<div class="parent"><div><span>父组件数据</span><input v-model="msg"> ' +
            '</div><p>{{msg}}</p><child :child-msg="msg"></child></div>',
        components: {
            'child': childNode
        },
        data(){
            return {
                'msg':'match'
            }
        }
    };
    new Vue({       //创建根实例
        el: '#example',
        components: {
          'parent': parentNode
        }
    })
</script>
</body>
</html>
```

在示例中,除初始值外,父组件的值无法更新到子组件中。

(2)定义一个计算属性,处理 props 的值并返回,代码如下:

```
props: ['size'],
computed: {
  normalizedSize: function () {
    return this.size.trim().toLowerCase()
  }
}
```

但是,由于是计算属性,因此只能显示值,不能设置值,示例代码如下:

```
<script>
    var childNode = {
            template:
        '<div class="child"><div><span>子组件数据</span>' +
            '<input v-model="temp"></div><p>{{temp}}</p></div>',
          props:['childMsg'],
          computed:{
              temp(){
                  return this.childMsg
              }
          },
    };
    var parentNode = {
            template:'<div class="parent"><div><span>父组件数据</span> ' +
            '<input v-model="msg"></div><p>{{msg}}</p>' +
            '<child :child-msg="msg"></child></div>',
        components: {
            'child': childNode
        },
    data(){
        return {
            'msg':'match'
        }
    }
    };
    new Vue({        //创建根实例
        el: '#example',
        components: {
            'parent': parentNode
        }
    })
</script>
```

示例中,由于子组件使用的是计算属性,所以子组件的数据无法手动修改。

(3)更加妥帖的方案是,使用变量储存 props 的初始值,并使用 watch 观察 props 的值的变化。当 props 的值发生变化时,更新变量的值,代码如下:

```html
<div id="example">
    <parent></parent>
</div>
<script>
    var childNode = {
        template:
        '<div class="child"><div><span>子组件数据</span>' +
            '<input v-model="temp"></div><p>{{temp}}</p></div>',
        props:['childMsg'],
        data(){
            return{
                temp:this.childMsg
            }
        },
        watch:{
            childMsg(){
                this.temp = this.childMsg
            }
        }
    };
    var parentNode = {
        template:
        '<div class="parent"><div><span>父组件数据</span>' +
            '<input v-model="msg"></div><p>{{msg}}</p>' +
            '<child :child-msg="msg"></child></div>',
        components: {
            'child': childNode
        },
        data(){
            return {
                'msg':'match'
            }
        }
    };
    new Vue({    //创建根实例
        el: '#example',
        components: {
            'parent': parentNode
        }
    })
</script>
```

5.4 组件通信

在Vue组件通信中其中最常见的通信方式就是父子组件之中的通信,而父子组件的设定方式在不同情况下又各不相同,归纳起来,组件之间通信如图5-8所示。最常见的就是父组件为控制组件而子组件为视图组件。父组件传递数据给子组件使用,遇到业务逻辑操作时子组件触发父组件的自定义事件。

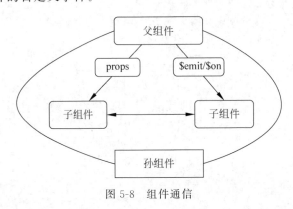

图5-8 组件通信

作为一个Vue初学者不得不了解的就是组件间的数据通信(暂且不谈Vuex,后面章节会讲到)。通信方式根据组件之间的关系有不同之处。组件关系有下面3种:父→子,子→父,非父子。

我们已经知道,父组件向子组件通信是通过props传递数据的,就好像方法的传参一样,父组件调用子组件并传入数据,子组件接收父组件传递的数据进行验证后使用。

props可以是数组也可以是对象,用于接收来自父组件的数据。props可以是简单的数组或者使用对象作为替代,对象允许配置高级选项,如类型检测、自定义校验和设置默认值。

5.4.1 自定义事件

当子组件需要向父组件传递数据时,就要用到自定义事件。v-on指令除了监听DOM事件外,还可以用于组件之间的自定义事件。

在子组件中用$emit()来触发事件以便将内部的数据传递给父组件,如例5-7所示。

【例5-7】 子组件向父组件传递数据

```
//第5章/子组件向父组件传递数据.html
<!DOCTYPE html>
<html>
<head>
<title></title>
```

```html
<meta charset="utf-8"/>
<script src="https://cdn.jsdelivr.net/npm/vue/dist/vue.js"></script>
</head>
<body>
<div id="app">
    <my-component v-on:myclick="onClick"></my-component>
</div>
<script>
    Vue.component('my-component', {
        template:'<div>' +
            '<button type="button" @click="childClick">单击触发自定义事件</button>' +
            '</div>',
        methods: {
            childClick () {
                this.$emit('myclick', '这是我传递出去的数据', '这是我传递出去的数据2')
            }
        }
    })
    new Vue({
        el: '#app',
        methods: {
            onClick () {
                console.log(arguments)
            }
        }
    })
</script>
</body>
</html>
```

解析上面示例中代码的执行步骤：
(1) 子组件在自己的方法中将自定义事件及需要传递的数据通过以下代码传递出去：

```
this.$emit('myclick','这是我传递出去的数据', '这是我传递出去的数据2')
```

- 第一个参数是自定义事件的名字。
- 后面的参数是依次想要传递出去的数据。

(2) 父组件利用 v-on 为事件绑定处理器，代码如下：

```
<my-component v-on:myclick="onClick"></my-component>
```

这样，在 Vue 实例的 methods 方法中就可以调用传进来的参数了。

5.4.2 $emit/$on

这种方法通过一个空的 Vue 实例作为中央事件中心,用它来触发事件和监听事件,巧妙而轻量地实现了任何组件间的通信,包括父子、兄弟、跨级。当我们的项目比较大时,可以选择更好的状态管理解决方案 Vuex。

实现方式如下:

```
var Event = new Vue();
    Event.$emit(事件名,数据);
    Event.$on(事件名,data => {});
```

假设兄弟组件有 3 个,分别是 A、B、C 组件,C 组件如何获取 A 或者 B 组件的数据,如例 5-8 所示。

【例 5-8】 组件之间的通信

```
//第5章/$emit-$on.html
<!DOCTYPE html>
<html>
<head>
<title></title>
<meta charset = "utf-8"/>
<script src = "https://cdn.jsdelivr.net/npm/vue/dist/vue.js"></script>
</head>
<body>
<div id = "app">
    <my-a></my-a>
    <my-b></my-b>
    <my-c></my-c>
</div>
<template id = "a">
    <div>
        <h3>A组件:{{name}}</h3>
        <button @click = "send">将数据发送给C组件</button>
    </div>
</template>
<template id = "b">
    <div>
        <h3>B组件:{{age}}</h3>
        <button @click = "send">将数组发送给C组件</button>
    </div>
</template>
<template id = "c">
    <div>
```

```
            <h3>C 组件:{{name}},{{age}}</h3>
        </div>
</template>
<script>
    var Event = new Vue();            //定义一个空的 Vue 实例
    var A = {
        template: '#a',
        data() {
        return {
            name: 'beixi'
        }
        },
        methods: {
            send() {
                Event.$emit('data-a', this.name);
            }
        }
    }
    var B = {
        template: '#b',
        data() {
        return {
            age: 18
        }
        },
        methods: {
            send() {
                Event.$emit('data-b', this.age);
            }
        }
    }
    var C = {
        template: '#c',
        data() {
        return {
            name: '',
            age: ""
        }
        },
        mounted() {                   //在模板编译完成后执行
            Event.$on('data-a', name => {
                this.name = name;     //箭头函数内部不会产生新的 this,此处如果不用 =>,
                                      //则 this 指代 Event
            })
            Event.$on('data-b', age => {
```

```
                this.age = age;
            })
        }
    }
    var vm = new Vue({
        el: '#app',
        components: {
            'my-a': A,
            'my-b': B,
            'my-c': C
        }
    });
</script>
</body>
</html>
```

页面显示效果如图5-9所示。

图5-9　组件间通信效果图

$on监听了自定义事件data-a和data-b,因为有时不确定何时会触发事件,一般会在mounted或created钩子中进行监听。

5.5　内容分发

在实际项目开发中,时常会把父组件的内容与子组件自己的模板混合起来使用。而这样的一个过程在Vue中被称为内容分发,也常常被称为slot(插槽)。其主要参照了当前Web Components规范草案,使用特殊的<slot>元素作为原始内容的插槽。

5.5.1　基础用法

由于slot是一块模板,因此对于任何一个组件,从模板种类的角度来分,其实都可分为非插槽模板和插槽模板。其中非插槽模板指的是HTML模板(也就是HTML的一些元素,例如div、span等构成的元素),其显示与否及怎样显示完全由插件自身控制,而插槽模板(也就是 slot)是一个空壳子,它显示与否及怎样显示完全由父组件来控制。不过,插槽显示的位置由子组件自身决定,slot 写在组件 template 的哪部分,父组件传过来的模板将来就显示在哪部分。

一般定义子组件的代码如下:

```
<div id="app">
    <child>
        <span>123456</span>
    </child>
</div>
<script>
    new Vue({
        el:'#app',
        components:{
            child:{
                template:"<div>这是子组件内容</div>"
            }
        }
    });
</script>
```

页面显示结果:这是子组件内容。<span>123456</span>内容并不会显示。

注意:虽然<span>标签被子组件的 child 标签所包含,但由于它不在子组件的 template 属性中,因此不属于子组件。

在 template 中添加<slot></slot>标签,代码如下:

```
<div id="app">
    <child>
        <span>123456</span>
    </child>
</div>
<script>
    new Vue({
        el:'#app',
        components:{
            child:{
                template:"<div><slot></slot>这是子组件内容</div>"
            }
        }
    });
</script>
```

页面显示结果:123456 这是子组件内容。

我们分步解析一下内容分发,现在我们看一个架空的例子,帮助理解刚刚讲过的严谨而难懂的定义。假设有一个组件名为 my-component，其使用上下文代码如下：

```
<my-component>
    <p>hi,slots</p>
</my-component>
```

再假设此组件的模板为：

```
<div>
    <slot></slot>
<div>
```

那么注入后的组件 HTML 相当于：

```
<div>
    <p>hi,slots</p>
<div>
```

标签<slot>会把组件使用上下文的内容注入此标签所占据的位置。组件分发的概念简单而强大，因为它意味着对一个隔离的组件除了通过属性、事件交互之外，还可以注入内容。

将此案例变成可以执行的代码，代码如下：

```
//部分代码省略
<div class="" id="app">
    <my-component>
        <p>hi,slots</p>
    </my-component>
</div>
<script>
    Vue.component('my-component', {
            template:'<div><slot></slot><div>'
    });
    new Vue({
        el:"#app"
    });
</script>
```

一个组件如果需要外部传入简单数据，如数字、字符串等，可以使用 property；如果需要传入 js 表达式或者对象，可以使用事件；如果希望传入的是 HTML 标签，那么使用内容分发就再好不过了。所以，尽管内容分发这个概念听起来极为复杂，而实际上可以简化为把 HTML 标签传入组件的一种方法。所以归根结底，内容分发是一种为组件传递参数的方法。

5.5.2 编译作用域

在深入了解分发 API 之前,先明确内容在哪个作用域里编译。假定模板为:

```html
<child-component>
  {{ message }}
</child-component>
```

这里的 message 应该绑定父组件的数据,还是绑定子组件的数据呢?答案是 message 就是一个 slot,但它绑定的是父组件的数据,而不是组件 <child-component> 的数据。

组件作用域简单地来说就是:父组件模板的内容在父组件作用域内编译;子组件模板的内容在子组件作用域内编译,如例 5-9 所示。

【例 5-9】 组件作用域

```html
//第5章/组件作用域.html
<!DOCTYPE html>
<html>
<head>
<title></title>
<meta charset="utf-8"/>
<script src="https://cdn.jsdelivr.net/npm/vue/dist/vue.js"></script>
</head>
<body>
<div id="app">
    <child-component v-show="someChildProperty"></child-component>
</div>
<script>
    Vue.component('child-component', {
        template: '<div>这是子组件内容</div>',
        data: function () {
            return {
                someChildProperty: true
            }
        }
    })
    new Vue({
        el:'#app'
    })
</script>
</body>
</html>
```

这里 someChildProperty 绑定的是父组件的数据,所以是无效的,因此获取不到数据。如果想在子组件上绑定,可以采用如下代码:

```
<div id="app">
    <child-component></child-component>
</div>
<script>
    Vue.component('child-component', {
        // 有效,因为是在正确的作用域内
        template: '<div v-show="someChildProperty">这是子组件内容</div>',
        data: function () {
            return {
                someChildProperty: true
            }
        }
    })
    new Vue({
        el:'#app'
    })
</script>
```

因此,slot 分发的内容是在父作用域内进行编译的。

5.5.3 默认 slot

如果要使父组件在子组件中插入内容,必须在子组件中声明 slot 标签,如果子组件模板不包含<slot>插口,父组件的内容将会被丢弃,如例 5-10 所示。

【例 5-10】 默认 slot

```
//第 5 章/默认 slot.html
<!DOCTYPE html>
<html>
<head>
<title></title>
<meta charset="utf-8"/>
<script src="https://cdn.jsdelivr.net/npm/vue/dist/vue.js"></script>
</head>
<body>
<div id="app">
    <!-- 1.2 组件 innerHTML 位置以后不管有什么代码,都会被放进插槽里面去 -->
    <index>
        <span>首页</span>
        <span>首页</span>
        <span>首页</span>
        <h1>手机</h1>
    </index>
</div>
```

```
<script>
    //插槽的作用就是将组件外部获取的代码片段放到组件内部
    /*定义默认插槽,通过slot组件定义,定义好了之后,就相当于一个插槽,你可以把它理解为
计算机的UBS插口*/
    Vue.component('index', {
        template:'<div>index</div>'
    })
    var vm = new Vue({
        el: '#app',
    })
</script>
</body>
</html>
```

页面显示结果为：index。所有子组件中的内容都不会被显示,而是被丢弃。要使父组件在子组件中插入内容,必须在子组件中声明slot标签,示例代码如下：

```
<script>
    Vue.component('index', {
        template:'<div><slot></slot>index</div>'
    })
    var vm = new Vue({
        el: '#app',
    })
</script>
```

5.5.4 具名slot

slot元素可以用一个特殊的属性name来配置如何分发内容。多个slot标签可以有不同的名字,如例5-11所示。

使用方法如下：

(1) 父组件要在分发的标签中添加属性"slot=name名"。

(2) 子组件在对应分发位置上的slot标签中添加属性"name=name名"。

【例5-11】 多个slot应用

```
//第5章/多个slot应用.html
<!DOCTYPE html>
<html>
<head>
<title></title>
<meta charset="utf-8"/>
<script src="https://cdn.jsdelivr.net/npm/vue/dist/vue.js"></script>
```

```
        </head>
        <body>
            <div id="app">
                <child>
                    <span slot="one">123456</span>
                    <span slot="two">abcdef</span>
                </child>
            </div>
            <script>
            new Vue({
                el:'#app',
                components:{
                    child:{
                        template:"<div><slot name='two'></slot>我是子组件<slot name='one'>
                        </slot></div>"
                    }
                }
            });
            </script>
        </body>
        </html>
```

页面显示结果为：abcdef 我是子组件 123456。

总结：slot 分发其实就是父组件在子组件内放一些 DOM，它负责这些 DOM 是否显示，以及在哪个地方显示。

5.5.5 作用域插槽

插槽分为单个插槽、具名插槽和作用域插槽，前两种比较简单并且在前面已将讲解过，本节的重点是讨论作用域插槽。

简单来说，前两种插槽的内容和样式皆由父组件决定，也就是说显示什么内容和怎样显示都由父组件决定，而作用域插槽的样式仍由父组件决定，但内容由子组件控制。即前两种插槽不能绑定数据，而作用域插槽是一个带绑定数据的插槽。

作用域插槽更具代表性的应用是列表组件，允许组件自定义如何渲染列表的每一项，代码如下：

```
<div id="app">
    <child></child>
</div>
<script>
    Vue.component('child', {
        data(){
```

```
            return {
                list:[1,2,3,4]
            }
        },
        template: '<div><ul>' +
            '<li v-for="item of list">{{item}}</li></ul></div>',
    })
    var vm = new Vue({
        el: '#app'
    })
</script>
```

在上面示例代码中,如果需要 child 组件在很多地方被调用,那么我们希望在不同的地方调用 child 的组件时,明确这个列表到底怎么循环。列表的样式不是由 child 组件控制的,而是由外部 child 模板占位符告诉我们组件的每一项该如何渲染,也就是说这里不应采用 li 标签,而是应该采用 slot 标签,示例代码如下:

```
<div id="app">
    <child>
        <template slot-scope="props"><!--固定写法,属性值可以自定义-->
            <li>{{props.item}}</li><!--插值表达式可以直接使用-->
        </template>
    </child>
</div>
<script>
    Vue.component('child', {
        data(){
            return {
                list:[1,2,3,4]
            }
        },
        template: '<div><ul>' +
            '<slot v-for="item of list" :item=item></slot></ul></div>',
    })
    var vm = new Vue({
        el: '#app'
    })
</script>
```

<slot v-for="item of list" :item=item></slot>这段代码的意思是 child 组件去实现一个列表的循环,但是列表项中的每一项怎样显示,并不用关心,具体怎样显示,由外部来决定。

<template slot-scope="props"></template>是一个固定写法,属性值可以自定义。

它的意思是当子组件用 slot 时，会向子组件传递一个 item，子组件接收的数据都放在 props 上。

什么时候使用作用域插槽呢？当子组件循环或某一部分的 DOM 结构应该由外部传递进来时，我们要用作用域插槽。使用作用域插槽，子组件可以向父组件的作用域插槽传递数据，父组件如果想接收这个数据，必须在外层使用 template 模版占位符，同时通过 slot-scope 对应的属性名字，接收传递过来的数据。如上面代码，传递一个 item 过来，在父组件的作用域插槽里就可以接收到这个 item，然后就可以使用它了。

5.6 动态组件

让多个组件使用同一个挂载点，并动态切换，这就是动态组件。通过使用保留的 <component> 元素，动态地绑定它的 is 特性，可以实现动态组件。它的应用场景往往应用在路由控制或者 tab 切换中。

5.6.1 基本用法

我们通过一个切换页面的例子来说明动态组件的基本用法，如例 5-12 所示。

【例 5-12】 切换页面

```
//第 5 章/切换页面.html
<!DOCTYPE html>
<html>
<head>
<title></title>
<meta charset="utf-8"/>
<script src="https://cdn.jsdelivr.net/npm/vue/dist/vue.js"></script>
</head>
<body>
<div id="app">
    <button @click="change">切换页面</button>
    <component :is="currentView"></component>
</div>
<script>
    new Vue({
        el:'#app',
        data:{
            index:0,
            arr:[
                {template:'<div>我是主页</div>'},
                {template:'<div>我是提交页</div>'},
                {template:'<div>我是存档页</div>'}
            ],
```

```
            },
            computed:{
                currentView(){
                    return this.arr[this.index];
                }
            },
            methods:{
                change(){
                    this.index = (++this.index) % 3;
                }
            }
        })
    </script>
</body>
</html>
```

component 标签中 is 属性决定了当前采用的子组件, :is 是 v-bind 的简写, 绑定了父组件中 data 的 currentView 属性。单击按钮时, 会更改数组 arr 的索引值, 同时也修改了子组件的内容。

5.6.2 keep-alive

动态切换掉的组件(非当前显示的组件)被移除掉了, 如果把切换出去的组件保留在内存中, 可以保留它的状态或避免重新渲染。<keep-alive>包裹动态组件时, 会缓存不活动的组件实例, 而不是销毁它们, 以便提高提取效率。和<transition>相似, <keep-alive>是一个抽象组件, 它自身不会渲染一个 DOM 元素, 也不会出现在父组件链中, 如例 5-13 所示。

【例 5-13】 keep-alive 基础用法

```
//第5章/keep-alive 基础用法.html
<!DOCTYPE html>
<html>
<head>
<title></title>
<meta charset = "utf-8"/>
<script src = "https://cdn.jsdelivr.net/npm/vue/dist/vue.js"></script>
</head>
<body>
<div id = "app">
    <button @click = "change">切换页面</button>
    <keep-alive>
        <component :is = "currentView"></component>
    </keep-alive>
</div>
```

```
<script>
    new Vue({
        el:'#app',
        data:{
            index:0,
            arr:[
                {template:'<div>我是主页</div>'},
                {template:'<div>我是提交页</div>'},
                {template:'<div>我是存档页</div>'}
            ],
        },
        computed:{
            currentView(){
                return this.arr[this.index];
            }
        },
        methods:{
            change(){
                /* ES6 新增了 let 命令,用来声明变量。它的用法类似于 var,但是所声明的变
量只在 let 命令所在的代码块内有效。*/
                let len = this.arr.length;
                this.index = (++this.index) % len;
            }
        }
    })
</script>
</body>
</html>
```

如果有多个条件性的子元素,<keep-alive>要求同时只有一个子元素被渲染时可以使用条件语句进行判断,如例 5-14 所示。

【例 5-14】 利用条件判断缓冲子元素

```
//第 5 章/利用条件判断缓冲子元素.html
<!DOCTYPE html>
<html>
<head>
<title></title>
<meta charset = "utf-8"/>
<script src = "https://cdn.jsdelivr.net/npm/vue/dist/vue.js"></script>
</head>
<body>
<div id = "app">
    <button @click = "change">切换页面</button>
```

```
            <keep-alive>
                <home v-if="index === 0"></home>
                <posts v-else-if="index === 1"></posts>
                <archive v-else></archive>
            </keep-alive>
        </div>
        <script>
            new Vue({
                el:'#app',
                components:{
                    home:{template:'<div>我是主页</div>'},
                    posts:{template:'<div>我是提交页</div>'},
                    archive:{template:'<div>我是存档页</div>'},
                },
                data:{
                    index:0,
                },
                methods:{
                    change(){
                        //在 data 外面定义的属性和方法通过 $options 可以获取和调用
                        let len = Object.keys(this.$options.components).length;
                        this.index = (++this.index) % len;
                    }
                }
            })
        </script>
    </body>
</html>
```

5.6.3 activated 钩子函数

Vue 给组件提供了 activated 钩子函数,作用于动态组件切换或者静态组件初始化的过程中。activated 是和 template、data 等属性平级的一个属性,其形式是一个函数,函数里默认有一个参数,而这个参数是一个函数,执行这个函数时,才会切换组件,即可延迟执行当前的组件,如例 5-15 所示。

【例 5-15】 activated 钩子函数

```
//第 5 章/activated 钩子函数.html
<!DOCTYPE html>
<html>
    <head>
        <title></title>
        <meta charset="utf-8"/>
```

```html
<script src="https://cdn.jsdelivr.net/npm/vue/dist/vue.js"></script>
</head>
<body>
<div id="app">
    <button @click='toShow'>单击显示子组件</button>
    <!----或者采用<component v-bind:is="which_to_show" keep-alive></component>也行----->
    <keep-alive>
        <component v-bind:is="which_to_show"></component>
    </keep-alive>
</div>
<script>
    var vm = new Vue({                  //创建根实例
        el: '#app',
        data: {
            which_to_show: "first"
        },
        methods: {
            toShow: function () {       //切换组件显示
                var arr = ["first", "second", "third", ""];
                var index = arr.indexOf(this.which_to_show);
                if (index < 2) {
                    this.which_to_show = arr[index + 1];
                } else {
                    this.which_to_show = arr[0];
                }
                console.log(this.$children);
            }
        },
        components: {
            first: {                    //第1个子组件
                template: "<div>这里是子组件1</div>"
            },
            second: {                   //第2个子组件
                template: "<div>这里是子组件2,这里是延迟后的内容: {{hello}}</div>",
                data: function () {
                    return {
                        hello: ""
                    }
                },
                activated: function (done) {           //执行这个参数时,才会切换组件
                    console.log('beixi')
                    var self = this;
                    var startTime = new Date().getTime();   //获得当前时间
                    //2s后执行
                    while (new Date().getTime() < startTime + 2000){
```

```
                    self.hello = '我是延迟后的内容';
                }
            }
        },
        third: { //第 3 个子组件
            template: "<div>这里是子组件 3</div>"
        }
    }
});
</script>
</body>
</html>
```

当切换到第 2 个组件的时候,会先执行 activated 钩子函数,会在 2s 后显示组件 2,起到了延迟加载的作用。

5.6.4 异步组件

在大型应用中,我们可能需要将应用分割成小一些的代码块,并且只在需要的时候才从服务器加载所需的模块。为了简化,Vue 允许以一个工厂函数的方式定义组件,这个工厂函数会异步解析所定义的组件。Vue 只有在这个组件需要被渲染的时候才会触发该工厂函数,且会把结果缓存起来供未来重新渲染,代码如下:

```
<div id="app">
    <async-example></async-example>
</div>
<script>
    Vue.component('async-example', function (resolve, reject) {
        setTimeout(function () {
            //向 resolve 回调传递组件定义
            resolve({
                template: '<div>这是异步渲染的内容!</div>'
            })
        }, 1000)
    })
    new Vue({
        el:'#app'
    })
</script>
```

由此所见,这个工厂函数会收到一个 resolve 回调,这个回调函数会在服务器得到组件定义的时候被调用。也可以调用 reject(reason) 来表示加载失败。这里的 setTimeout 是为了演示异步,如何获取组件取决于自己。例如把组件配置成一个对象,通过 Ajax 进行请求,然后调用 reslove 传入配置选项。

5.6.5　ref 和 $refs

在 Vue 中一般很少直接操作 DOM,但不可避免有时候确定需要用到,这时我们可以通过 ref 和 $refs 来实现:

(1) ref:ref 被用来给元素或子组件注册引用信息,引用信息将会注册在父组件的 $refs 对象上,如果在普通的 DOM 元素上使用,引用指向的就是 DOM 元素,如果在子组件上,引用指向的就是组件的实例。

(2) $refs:$refs 是一个对象,持有已注册过 ref 的所有子组件。

1. 普通获取 DOM 的方式

我们先通过 getElementById 方法获取,代码如下:

```
<div id="app">
    <input type="button" value="获取 h3 的值" @click="getElement">
    <h3 id="myh3">我是 h3</h3>
</div>
<script>
    var vm = new Vue({
        el:"#app",
        data:{},
        methods:{
            getElement(){
                //通过 getElementById 方式获取 DOM 对象
                console.log(document.getElementById("myh3").innerHTML);
            }
        }
    })
</script>
```

2. ref 使用

接下来我们通过 ref 属性来获取,代码如下:

```
<div id="app">
    <input type="button" value="获取 h3 的值" @click="getElement">
    <h3 id="myh3" ref="myh3">我是 h3</h3>
</div>
```

然后在控制台查看 vm 实例对象,如图 5-10 所示。

通过上面的示例我们发现,在 vm 实例上有一个 $refs 属性,而且该属性拥有我们通过 ref 注册的 DOM 对象,于是我们可以这样获取 DOM 对象,代码如下:

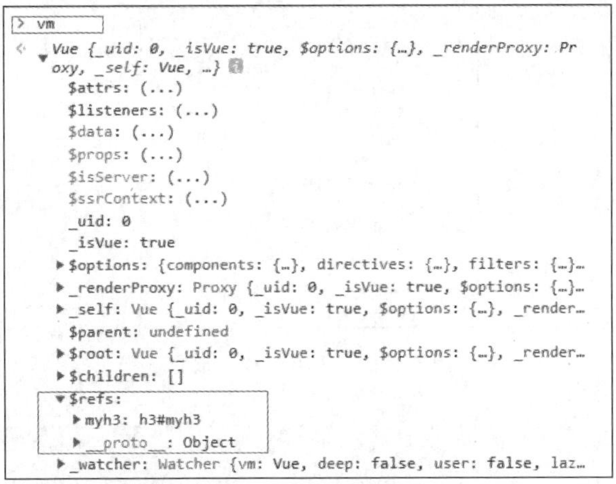

```
> vm
< Vue {_uid: 0, _isVue: true, $options: {…}, _renderProxy: Pr
   oxy, _self: Vue, …}
     $attrs: (...)
     $listeners: (...)
     $data: (...)
     $props: (...)
     $isServer: (...)
     $ssrContext: (...)
     _uid: 0
     _isVue: true
   ▶ $options: {components: {…}, directives: {…}, filters: {…}…
   ▶ _renderProxy: Proxy {_uid: 0, _isVue: true, $options: {…}…
   ▶ _self: Vue {_uid: 0, _isVue: true, $options: {…}, _render…
     $parent: undefined
   ▶ $root: Vue {_uid: 0, _isVue: true, $options: {…}, _render…
   ▶ $children: []
   ▼ $refs:
     ▶ myh3: h3#myh3
     ▶ __proto__: Object
   ▶ _watcher: Watcher {vm: Vue, deep: false, user: false, laz…
```

图 5-10 vm 实例对象信息

```
<script>
    var vm = new Vue({
        el:"#app",
        data:{},
        methods:{
            getElement(){
                console.log(this.$refs.myh3.innerHTML);
            }
        }
    })
</script>
```

3. ref 在组件中使用

在子组件中使用 ref 属性,会将子组件添加到父组件的 $refs 对象中,代码如下:

```
<div id="app">
    <input type="button" value="获取 h3 的值" @click="getElement">
    <h3 id="myh3" ref="myh3">我是 h3</h3>
    <hr>
    <login ref="mylogin"></login>
</div>
```

在控制台输入 vm,查看 vm 对象,如图 5-11 所示。

通过 vm 实例查看,发现 $refs 中已绑定 mylogin 组件,而且还看到了对应组件中的 msg 属性和 show 方法,这样我们便可以调用它们了,代码如下:

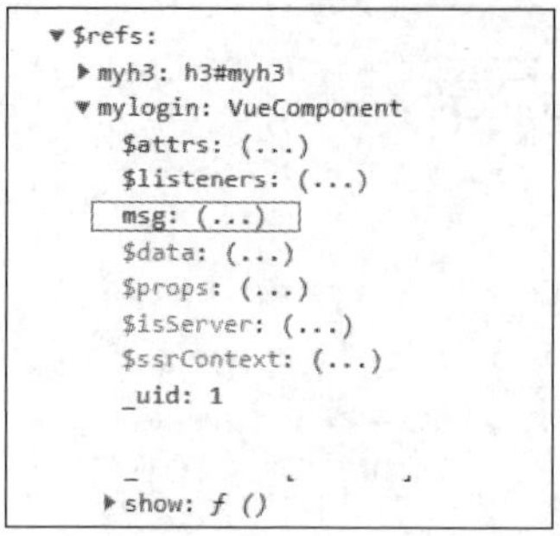

图 5-11　vm 实例对象信息

```
var vm = new Vue({
    el:"#app",
    data:{},
    methods:{
        getElement(){
            //通过 getElementById 方式获取 DOM 对象
            //console.log(this.$refs.myh3.innerHTML);
            console.log(this.$refs.mylogin.msg);
            this.$refs.mylogin.show();
        }
    },
    components:{
        login
    }
})
```

完整实例代码如例 5-16 所示。

【例 5-16】　ref 在组件中的应用

```
//第 5 章/ref 在组件中的应用.html
<!DOCTYPE html>
<html lang="en">
<head>
<meta charset="UTF-8">
<title>Document</title>
<!-- 引入 vue -->
<script src="https://cdn.jsdelivr.net/npm/vue/dist/vue.js"></script>
</head>
<body>
<div id="app">
    <input type="button" value="获取 h3 的值" @click="getElement">
    <h3 id="myh3" ref="myh3">我是 h3</h3>
```

```html
            <hr>
            <login ref="mylogin"></login>
</div>
<script>
    var login = {
        template: "<h3>我是login子组件</h3>",
        data(){
            return {
                msg: "ok"
            }
        },
        methods:{
            show(){
                console.log("show方法执行了...")
            }
        }
    }
    var vm = new Vue({
        el:"#app",
        data:{},
        methods:{
            getElement(){
                //通过getElementById方式获取DOM对象
                //console.log(this.$refs.myh3.innerHTML);
                console.log(this.$refs.mylogin.msg);
                this.$refs.mylogin.show();
            }
        },
        components:{
            login
        }
    })
</script>
</body>
</html>
```

5.7 综合案例

添加组件后,我们通过单击"发表评论"按钮来添加内容并传递到评论列表中,如例5-17所示。实现的逻辑是:

(1) 通过单击"发表评论"按钮触发单击事件并调用组件中methods所定义的方法。
(2) 在methods所定义的方法中加载并保存localStorage的列表数据到list中。
(3) 将录入的信息添加到list中,然后将数据保存到localStorage中。

(4) 调用父组件中的方法来刷新列表数据。

【例 5-17】 综合案例

```html
//第 5 章/综合案例.html
<!DOCTYPE html>
<html lang="en">
<head>
    <meta charset="UTF-8">
    <meta name="viewport" content="width=device-width, initial-scale=1.0">
    <meta http-equiv="X-UA-Compatible" content="ie=edge">
    <title>Document</title>
    <!-- 引入 vue -->
    <script src="https://cdn.jsdelivr.net/npm/vue/dist/vue.js"></script>
    <!-- 引入 bootstrap -->
    <link rel="stylesheet" href="./lib/bootstrap-3.3.7.css">
</head>
<body>
<div id="app">
    <cmt-box @func="loadComments"></cmt-box>
    <ul class="list-group">
        <li class="list-group-item" v-for="item in list" :key="item.id">
            <span class="badge">评论人：{{ item.user }}</span>
            {{ item.content }}
        </li>
    </ul>
</div>
<template id="tmpl">
    <div>
        <div class="form-group">
            <label>评论人：</label>
            <input type="text" class="form-control" v-model="user">
        </div>
        <div class="form-group">
            <label>评论内容：</label>
            <textarea class="form-control" v-model="content"></textarea>
        </div>
        <div class="form-group">
            <input type="button" value="发表评论" class="btn btn-primary" @click="postComment">
        </div>
    </div>
</template>
<script>
    var commentBox = {
        data() {
            return {
```

```
                    user: '',
                    content: ''
                }
            },
            template: '#tmpl',
            methods: {
                postComment() {        //发表评论的方法
                    var comment = { id: Date.now(), user: this.user, content: this.content }
                    // 从 localStorage 中获取所有的评论
                    var list = JSON.parse(localStorage.getItem('cmts') || '[]')
                    list.unshift(comment)
                    // 重新保存最新的评论数据
                    localStorage.setItem('cmts', JSON.stringify(list))
                    this.user = this.content = ''
                    this.$emit('func')
                }
            }
        }
        //创建 Vue 实例,得到 ViewModel
        var vm = new Vue({
            el: '#app',
            data: {
                list: [
                    { id: Date.now(), user: 'beixi', content: '这是我的网名' },
                    { id: Date.now(), user: 'jzj', content: '这是我的真名' },
                    { id: Date.now(), user: '贝西奇谈', content: '有任何问题可以关注公众号' }
                ]
            },
            beforeCreate(){ /* 注意:这里不能调用 loadComments 方法,因为在执行这个钩子函数的时候,data 和 methods 都还没有被初始化*/
            },
            created(){
                this.loadComments()
            },
            methods: {
                loadComments() {       //从本地的 localStorage 中加载评论列表
                    var list = JSON.parse(localStorage.getItem('cmts') || '[]')
                    this.list = list
                }
            },
            components: {
                'cmt-box': commentBox
            }
        });
    </script>
</body>
</html>
```

页面显示效果如图 5-12 所示。

图 5-12　综合案例实现效果

第 6 章 过渡与动画

过渡是 Vue 为 DOM 动画效果提供的一个特性，Vue 在插入、更新或者移除 DOM 时，提供多种不同方式的应用过渡效果。包括以下工具：

(1) 在 CSS 过渡和动画中自动应用 class。
(2) 可以配合使用第三方 CSS 动画库，如 Animate.css。
(3) 在过渡钩子函数中使用 JavaScript 直接操作 DOM。
(4) 可以配合使用第三方 JavaScript 动画库，如 Velocity.js。

6.1 元素/组件过渡

Vue 提供了 transition 的封装组件，在下列情形中，可以给任何元素和组件添加进入/离开过渡。

(1) 条件渲染（使用 v-if）。
(2) 条件展示（使用 v-show）。
(3) 动态组件。
(4) 组件根节点。

语法格式如下：

```
<transition name = "nameoftransition">
    <div></div>
</transition>
```

我们可以通过实例 6-1 简单理解 Vue 的过渡是如何实现的。

【例 6-1】 组件过渡应用

```
//第 6 章/组件过渡应用.html
<!DOCTYPE html>
<html lang = "en">
<head>
```

```html
<meta charset="UTF-8">
<title>Document</title>
<!-- 引入 vue -->
<script src="https://cdn.jsdelivr.net/npm/vue/dist/vue.js"></script>
</head>
<body>
<div id="demo">
    <button v-on:click="show = !show">
        点我
    </button>
    <transition name="fade">
        <p v-if="show">动画实例</p>
    </transition>
</div>
<script>
    new Vue({
        el: '#demo',
        data: {
            show: true
        }
    })
</script>
</body>
</html>
```

实例中通过单击"点我"按钮将变量 show 的值从 true 变为 false。如果值为 true,则显示子元素 p 标签的内容。

下面这段代码展示了 transition 标签包裹了 p 标签,代码如下:

```html
<transition name="fade">
    <p v-if="show">动画实例</p>
</transition>
```

当插入或删除包含在 transition 组件中的元素时,Vue 将会进行以下处理:

(1) 自动嗅探目标元素是否应用了 CSS 过渡或动画。如果是,则在恰当的时机添加/删除 CSS 类名。

(2) 如果过渡组件提供了 JavaScript 钩子函数,这些钩子函数将在恰当的时机被调用。

(3) 如果没有找到 JavaScript 钩子函数并且也没有检测到 CSS 过渡/动画,DOM 操作(插入/删除)在下一帧中立即执行。(注意:此处指浏览器逐帧动画机制,和 Vue 的 nextTick 概念不同。)

6.2 使用过渡类实现动画

过渡其实就是一个淡入淡出的效果。Vue 在元素显示与隐藏的过渡中提供了以下 6 个 class 进行切换。

(1) v-enter：定义进入过渡的开始状态。在元素被插入之前生效，在元素被插入之后的下一帧移除。

(2) v-enter-active：定义进入过渡生效时的状态。在整个进入过渡的阶段中应用，在元素被插入之前生效，在过渡/动画完成之后移除。这个类可以被用来定义进入过渡的过程时间、延迟和曲线函数。

(3) v-enter-to：Vue 2.1.8 版及以上定义进入过渡的结束状态。在元素被插入之后下一帧生效（与此同时 v-enter 被移除），在过渡/动画完成之后移除。

(4) v-leave：定义离开过渡的开始状态。在离开过渡被触发时立刻生效，下一帧被移除。

(5) v-leave-active：定义离开过渡生效时的状态。在整个离开过渡的阶段中应用，在离开过渡被触发时立刻生效，在过渡/动画完成之后移除。这个类可以被用来定义离开过渡的过程时间、延迟和曲线函数。

(6) v-leave-to：Vue 2.1.8 版及以上定义离开过渡的结束状态。在离开过渡被触发之后下一帧生效（与此同时 v-leave 被删除），在过渡/动画完成之后移除。

原理如图 6-1 所示。

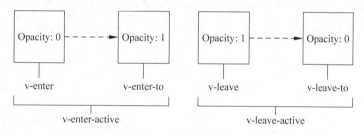

图 6-1　过渡类原理

对于这些在过渡中切换的类名来说，如果使用一个没有名字的 <transition>，则 v- 是这些类名的默认前缀。如果使用了 <transition name="my-transition">，那么 v-enter 会被替换为 my-transition-enter。

v-enter-active 和 v-leave-active 可以控制进入/离开过渡的不同的缓和曲线，在下面的章节中会有个示例说明。

6.2.1　CSS 过渡

常用的 Vue 过渡效果使用 CSS 过渡 transition。我们首先看一个简单的切换显示按钮，代码如下：

```
<div id="app">
    <button @click="show=!show">点我</button>
    <div v-if="show">Hello world</div>
</div>
<script>
    new Vue({
        el:'#app',
        data:{
            show:true
        }
    })
</script>
```

我们现在希望 Hello world 能有一个渐隐渐现的效果，那么就需要在 div 外层包裹一个 transition 标签，代码如下：

```
<div id="app">
    <button @click="show=!show">点我</button>
    <transition name="fade">
        <div v-if="show">Hello world</div>
    </transition>
</div>
```

当然只这样执行代码还无法形成过渡效果。仍需在 transition 标签中添加 class 样式，只不过 class 样式不需要我们手动添加，Vue 在运行中会自动地构建一个动画的流程，如图 6-2 所示。

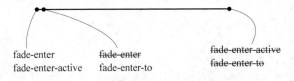

图 6-2　Vue 动画过渡图

当动画执行的一瞬间，会自动在 div 上增加两个 class 名字，分别是 fade-enter 和 fade-enter-active，因为 transition 标签的 name 是 fade。

当动画运行到第 2 帧时，Vue 又会把 fade-enter 删除，然后添加一个 fade-enter-to，再当动画执行到结束的一瞬间，Vue 又把 fade-enter-active 和 fade-enter-to 删除掉。

动画原理是不是这样，可以看一看下面的示例，代码如下：

```
<head>
<style>
```

```
            .fade-enter{
                opacity: 0;
            }
            .fade-enter-active{
                transition: opacity 3s;
            }
        </style>
    </head>
    <body>
        <div id="app">
            <button @click="show=!show">点我</button>
            <transition name="fade">
                <div v-if="show">Hello world</div>
            </transition>
        </div>
        <script>
            new Vue({
                el: '#app',
                data: {
                    show: true
                }
            })
        </script>
    </body>
```

上面示例已经有明显的过渡效果了。

在图 6-2 所示的 Vue 动画过渡图中，我们发现 fade-enter-active 是全程存在的，它的作用是，如果监听到了元素 opacity 发生了变化，那么就让这个变化在 3s 内完成。

fade-enter 在第 1 帧的时候存在，在第 2 帧的时候被删除。首先在第 1 帧的时候，fade-enter-active 和 fade-enter 同时存在，并且 opacity＝0，在第 2 帧的时候，fade-enter 被删除，opacity 恢复到原来的初始状态，也就是 1，在这个过程中，opacity 发生了变化，所以 fade-enter-active 就让这个变化在 3s 内完成。

如果在 transition 标签中不添加 name 属性，默认自动会加 name＝"v"，那么 class 样式应该如下面代码所示：

```
<style>
    .v-enter{
        opacity: 0;
    }
    .v-enter-active{
        transition: opacity 3s;
    }
</style>
```

再来看 Vue 元素从显示到隐藏的动画效果，代码如下：

```html
<style>
    .v-enter{
        opacity: 0;
    }
    .v-enter-active{
        transition: opacity 3s;
    }
    .v-leave-to{
        opacity: 0;
    }
    .v-leave-active{
        transition: opacity 3s;
    }
</style>
```

和上面 enter 的原理基本一致。

完整示例代码，如例 6-2 所示。

【例 6-2】 CSS 过渡

```html
//第 6 章/CSS 过渡.html
<!DOCTYPE html>
<html lang="en">
<head>
<meta charset="UTF-8">
<title>Document</title>
<!-- 引入 vue -->
<script src="https://cdn.jsdelivr.net/npm/vue/dist/vue.js"></script>
<style>
        .v-enter{
            opacity: 0;
        }
        .v-enter-active{
            transition: opacity 3s;
        }
        .v-leave-to{
            opacity: 0;
        }
        .v-leave-active{
            transition: opacity 3s;
        }
</style>
</head>
<body>
```

```
<div id="app">
    <button @click="show=!show">点我</button>
    <transition>
        <div v-if="show">Hello world</div>
    </transition>
</div>
<script>
    new Vue({
        el:'#app',
        data:{
            show:true
        }
    })
</script>
</body>
</html>
```

6.2.2　CSS 动画

CSS 动画 animation 用法与 CSS 过渡 transition 用法相同，区别是在动画中 v-enter 类名在节点插入 DOM 后不会立即被删除，而是在 animationend 事件触发时被删除，如例 6-3 所示。

【例 6-3】　在元素 enter 和 leave 时都增加缩放 scale 效果

```
//第6章/CSS动画.html
<!DOCTYPE html>
<html lang="en">
<head>
<meta charset="UTF-8">
<title>Document</title>
<!--引入vue-->
<script src="https://cdn.jsdelivr.net/npm/vue/dist/vue.js"></script>
<style>
    .bounce-enter-active{
        animation:bounce-in .5s;
    }
    .bounce-leave-active{
        animation:bounce-in .5s reverse;
    }
    @keyframes bounce-in{
        0%{transform:scale(0);}
        50%{transform:scale(1.5);}
        100%{transform:scale(1);}
```

```html
        }
    </style>
</head>
<body>
<div id="app">
    <button v-on:click="show = !show">点我</button>
    <transition name="bounce">
        <p v-if="show">Hello World</p>
    </transition>
</div>
<script>
    new Vue({
        el:'#app',
        data:{
            show:true
        }
    })
</script>
</body>
</html>
```

Vue 为了检测到过渡的完成,必须设置相应的事件监听器。它可以是 transitionend 或 animationend,这取决于元素所采用的 CSS 规则。如果使用其中任何一种,则 Vue 能自动识别类型并设置监听。

但是,在一些场景中,需要将一个元素同时设置两种过渡动效,例如 animation 很快地被触发并完成了,而 transition 效果还没结束。在这种情况下,就需要使用 type 特性并设置 animation 或 transition 来明确声明需要 Vue 监听的类型,如例 6-4 所示。

【例 6-4】 CSS 过渡和动画同时使用

```html
//第 6 章/CSS 过渡和动画同时使用.html
<!DOCTYPE html>
<html lang="en">
<head>
    <meta charset="UTF-8">
    <title>Document</title>
    <!-- 引入 vue -->
    <script src="https://cdn.jsdelivr.net/npm/vue/dist/vue.js"></script>
    <style>
        .fade-enter,.fade-leave-to{
            opacity:0;
        }
        .fade-enter-active,.fade-leave-active{
            transition:opacity 1s;
```

```
            animation:bounce - in 5s;
        }
        @keyframes bounce - in{
            0 % {transform:scale(0);}
            50 % {transform:scale(1.5);}
            100 % {transform:scale(1);}
        }
    </style>
</head>
<body>
<div id = "app">
    <button v - on:click = "show = !show">点我</button>
    <transition name = "fade" type = "transition">
        <p v - if = "show">Hello World</p>
    </transition>
</div>
<script>
    new Vue({
        el: '#app',
        data: {
            show: true,
        },
    })
</script>
</body>
</html>
```

6.2.3 自定义过渡的类名

我们可以通过以下特性来自定义过渡类名：

（1）enter-class。

（2）enter-active-class。

（3）leave-class。

（4）leave-active-class。

它们的优先级高于普通的类名，这对于 Vue 的过渡系统和其他第三方 CSS 动画库来讲，如与 Animate.css 结合使用十分有用，如例 6-5 所示。

【例 6-5】 自定义过渡类应用

```
//第6章/自定义过渡类应用.html
<!DOCTYPE html >
< html lang = "en">
< head >
```

```html
<meta charset="UTF-8">
<title>Document</title>
<!--引入vue-->
<script src="https://cdn.jsdelivr.net/npm/vue/dist/vue.js"></script>
<style>
        .fade-in-active,.fade-out-active{
            transition: all 1.5s ease
        }
        .fade-in-enter,.fade-out-active{
            opacity: 0
        }
</style>
</head>
<body>
<div id="app">
    <button @click="show = !show">
        点我
    </button>
    <transition
            name="fade"
            enter-class="fade-in-enter"
            enter-active-class="fade-in-active"
            leave-class="fade-out-enter"
            leave-active-class="fade-out-active">
        <p v-show="show">hello</p>
    </transition>
</div>
<script>
    new Vue({
        el: '#app',
        data: {
            show: true
        }
    })
</script>
</body>
</html>
```

上面代码中，原来默认的 fade-enter 类对应 fade-in-enter，而 fade-enter-active 类对应 fade-in-active，依次类推。

6.2.4　CSS 过渡钩子函数

除了可以用 CSS 过渡的动画来实现 Vue 的组件过渡，还可以用 JavaScript 的钩子函数来实现，在钩子函数中直接操作 DOM。可以在属性中声明 JavaScript 钩子，代码如下：

```html
<transition v-on:before-enter="beforeEnter" v-on:enter="enter" v-on:after-enter="afterEnter" v-on:enter-cancelled="enterCancelled" v-on:before-leave="beforeLeave" v-on:leave="leave" v-on:after-leave="afterLeave" v-on:leave-cancelled="leaveCancelled"><!-- ... -->
</transition>
```

JS 代码块如下：

```
//...
methods: {
    //--------
    //进入中
    //--------
    beforeEnter: function (el) {
        //...
    },
    //当与 CSS 结合使用时
    //回调函数 done 是可选的
    enter: function (el, done) {
        //...
        done()
    },
    afterEnter: function (el) {
        //...
    },
    enterCancelled: function (el) {
        //...
    },

    //--------
    //离开时
    //--------

    beforeLeave: function (el) {
        //...
    },
    //当与 CSS 结合使用时
    //回调函数 done 是可选的
    leave: function (el, done) {
        //...
        done()
    },
    afterLeave: function (el) {
        //...
    },
```

```
        //leaveCancelled 只用于 v-show 中
leaveCancelled: function (el) {
        //...
    }
}
```

这些钩子函数可以结合 CSS transitions/animations 使用，也可以单独使用。

注意：

（1）当只用 JavaScript 过渡的时候，在 enter 和 leave 中必须使用 done 进行回调。否则它们将被同步调用，过渡会立即完成。

（2）推荐对于仅使用 JavaScript 过渡的元素添加 v-bind:css="false"，这样 Vue 便会跳过 CSS 的检测。这也可以避免过渡过程中 CSS 的影响。

Vue.js 也可以和一些 JavaScript 动画库配合使用，这里只需调用 JavaScript 钩子函数，而不需要定义 CSS 样式。transition 接受选项 css.false，将直接跳过 CSS 检测，避免 CSS 规则干扰过渡，而且需要在 enter 和 leave 钩子函数中调用 done 函数，明确过渡结束时间。此处将引入 Velocity.js 配合使用 JavaScript 过渡，如例 6-6 所示。

【例 6-6】 一个使用 Velocity.js 的简单例子。

```
//第6章/CSS过渡钩子函数.html
<!DOCTYPE html>
<html lang="en">
<head>
<meta charset="UTF-8">
<title>Document</title>
<!-- 引入 vue -->
<script src="https://cdn.jsdelivr.net/npm/vue/dist/vue.js"></script>
<!--
Velocity 和 jQuery.animate 的工作方式类似，也是用来实现 JavaScript 动画的一个很棒的选择-->
<script src="https://cdnjs.cloudflare.com/ajax/libs/velocity/1.2.3/velocity.min.js">
</script>
</head>
<body>
<div id="app">
    <button @click="show = !show">
        点我
    </button>
    <transition
        v-on:before-enter="beforeEnter"
        v-on:enter="enter"
        v-on:leave="leave"
        v-bind:css="false">
        <p v-if="show">
```

```html
            Hello
        </p>
    </transition>
</div>
<script>
    new Vue({
        el: '#app',
        data: {
            show: false
        },
        methods: {
            beforeEnter: function (el) {
                el.style.opacity = 0
                el.style.transformOrigin = 'left'
            },
            enter: function (el, done) {
                Velocity(el, { opacity: 1, fontSize: '1.4em' }, { duration: 300 })
                Velocity(el, { fontSize: '1em' }, { complete: done })
            },
            leave: function (el, done) {
                Velocity(el, { translateX: '15px', rotateZ: '50deg' }, { duration: 600 })
                Velocity(el, { rotateZ: '100deg' }, { loop: 2 })
                Velocity(el, {
                    rotateZ: '45deg',
                    translateY: '30px',
                    translateX: '30px',
                    opacity: 0
                }, { complete: done })
            }
        }
    })
</script>
</body>
</html>
```

第 7 章 前端工程化

高效的开发离不开基础工程的搭建。本章主要介绍如何使用 Vue 进行实际 SPA 项目的开发，这里使用的是目前热门的 JavaScript 应用程序模块打包工具 Webpack，进行模块化开发、代码编译和打包。

7.1 Vue-cli

Vue 脚手架指的是 Vue-cli，它是一个专门为单页面应用快速搭建繁杂程序的脚手架，它可以轻松地创建新的应用程序而且可用于自动生成 Vue 和 Webpack 的项目模板。

Vue-cli 是一个基于 Vue.js 进行快速开发的完整系统，其提供以下功能：

（1）通过 @vue/cli 实现交互式的项目脚手架。

（2）通过 @vue/cli+@vue/cli-service-global 实现零配置原型开发。

（3）一个运行时的依赖（@vue/cli-service），该依赖：

- 可升级。
- 基于 Webpack 构建，并带有合理的默认配置。
- 可以通过项目内的配置文件进行配置。
- 可以通过插件进行扩展。

（4）一个丰富的官方插件集合，集成了前端生态中最好的工具。

（5）一套完全图形化的创建和管理 Vue.js 项目的用户界面。

Vue-cli 致力于将 Vue 生态中的工具基础标准化。它确保各种构建工具能够基于智能的默认配置即可平稳衔接，这样开发者可以专注在撰写应用上，而不必花费好几天时间去纠结配置的问题。与此同时，它也为每个工具提供了调整配置的灵活性，而无须 eject。

利用 Vue-cli 脚手架构建 Vue 项目需要先安装 Node.js 和 NPM 环境。

7.1.1 Node.js

1. 什么是 Node.js

Node.js 是一个基于 Chrome V8 引擎的 JavaScript 运行环境。Node.js 使用了一个事

件驱动、非阻塞式 I/O 的模型。

Node 是一个让 JavaScript 运行在服务端的开发平台，它让 JavaScript 成为与 PHP、Python、Perl、Ruby 等服务端语言平起平坐的脚本语言。Node 发布于 2009 年 5 月，由 Ryan Dahl 开发，实质是对 Chrome V8 引擎进行了封装。

Node 对一些特殊用例进行优化，提供替代的 API，使得 V8 在非浏览器环境下运行得更好。V8 引擎执行 JavaScript 的速度非常快，性能非常好。Node 是一个基于 Chrome JavaScript 运行时建立的平台，用于方便地搭建响应速度快、易于扩展的网络应用。Node 使用事件驱动，采用非阻塞 I/O 模型而得以轻量和高效，非常适合在分布式设备上运行数据密集型的实时应用。

2．Node.js 的安装

Node.js 的安装比较简单，大家需要在 Node.js 官网(https://nodejs.org/en/download/)下载并安装 Node.js 环境，Windows 系统推荐下载 Windows Installer (.msi)。同时，大家会得到一个附送的 NPM 工具。

(1) 安装 Node.js，双击下载好的 node 安装文件，如图 7-1 所示。
安装过程比较简单，一直单击"下一步"按钮即可。

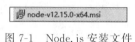

图 7-1　Node.js 安装文件

(2) 环境变量配置。安装完成后需要设置环境变量，即在 Path 中添加安装目录(例如：D:\java\nodejs)，如图 7-2 所示。

图 7-2　Node.js 环境变量配置

(3）单击"开始"按钮，然后单击"运行"按钮，在输入框内输入 cmd 命令，最后输入 node -v，如图 7-3 所示，验证安装是否成功。

图 7-3　验证 Node.js 安装是否成功

7.1.2　NPM

1．什么是 NPM

NPM 代表 npmjs.org 这个网站，这个站点存储了很多 Node.js 的第三方功能包。

NPM 的全称是 Node Package Manager，它是一个 Node.js 包管理和分发工具，已经成为非官方的发布 Node 模块（包）的标准。它可以让 JavaScript 开发者能够更加轻松地共享代码和共用代码片段，并且通过 NPM 管理需要分享的代码也很方便、快捷和简单。

NPM 最初用于管理和分发 Node.js 的依赖，它自动化的机制使得层层嵌套的依赖管理变得十分简单，因此后来被广泛应用于前端依赖的管理中。

NPM 是随同 Node.js 一起安装的包管理工具，能解决 Node.js 代码部署上的很多问题，常见的使用场景有以下几种：

（1）允许用户从 NPM 服务器下载别人编写的第三方包到本地使用。

（2）允许用户从 NPM 服务器下载并安装别人编写的命令行程序到本地使用。

（3）允许用户将自己编写的包或命令行程序上传到 NPM 服务器供别人使用。

2．NPM 安装

由于 Node.js 已经集成了 NPM，所以在安装 Node.js 时 NPM 也一并安装好了。所以可以在 cmd 终端输入 npm -v 来测试是否安装成功。命令如图 7-4 所示，如果出现版本提示则表示安装成功。

图 7-4　验证 NPM 安装是否成功

温馨提示：由于 NPM 的仓库源部署在国外，资源传输速度较慢且可能受限制，所以大家可以直接使用 NPM 安装其他依赖，也可以使用淘宝的镜像源 CNPM（CNPM 是淘宝团队在国内开发的一个相当于 NPM 的镜像，可以用 CNPM 代替 NPM 来安装依赖包）。

安装 CNPM 命令，我们通过打开命令行工具（Win＋R），输入：

```
npm install -g cnpm --registry=http://registry.npm.taobao.org
```

这样就可以使用淘宝定制的 CNPM(gzip 压缩支持) 命令行工具代替默认的 NPM，当然大家也可以选择不安装，不代替 NPM。

7.1.3 基本使用

在用 Vue 构建大型应用时推荐使用 NPM 安装。NPM 能很好地和诸如 WebPack 或 Browserify 模块打包器配合使用。在 7.1.1 节中将 Vue 环境搭建完成后，本节可以利用 Vue 提供的 Vue-cli 脚手架快速建立项目，步骤如下：

(1) 搭建第一个完整的 Vue-cli 脚手架构建的项目。

Vue-cli 是用 Node 编写的命令行工具，我们需要进行全局安装。打开命令行终端，输入如下命令：

```
npm install -g vue-cli
```

然后等待安装完成，如图 7-5 所示。

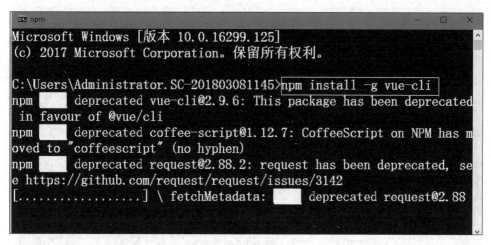

图 7-5　安装脚手架 Vue-cli

注意：只需第一次构建脚手架项目时执行上述命令，以后的 Vue 项目构建从第(3)步开始。

(2) 安装完成后，在命令行输入 vue -V，如果出现相应的版本号，则说明安装成功，如图 7-6 所示。

(3) 我们可以使用 Vue-cli 快速生成一个基于 Webpack 模板构建的项目，如图 7-7 所示，项目名为 vue-project。

首先需要在命令行中进入项目目录，然后输入如下命令：

```
//cd $自定义路径$ ( 如：D:\Web前端\VueCode)
vue init webpack vue-project
```

图 7-6　Vue-cli 安装成功

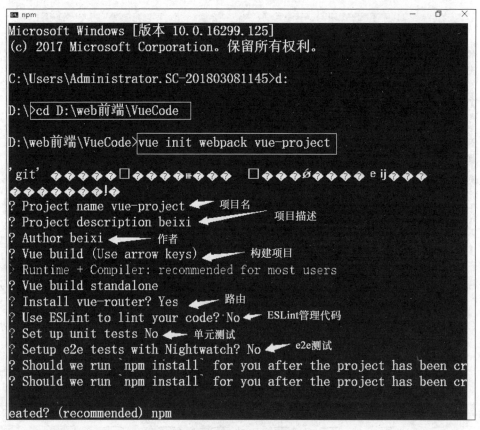

图 7-7　使用 Webpack 模板创建项目

(4) 配置完成后,可以看到目录下多出了一个项目文件夹,里面就是 Vue-cli 所创建的一个基于 Webpack 的 Vue.js 项目。

然后进入项目目录(如: cd　vue-project),使用 npm install 安装依赖,如图 7-8 所示。

图 7-8　安装依赖

依赖安装完成后,我们来看一下项目的目录结构,代码如下:

```
|-- build                        // 项目构建(Webpack)相关代码
|   |-- build.js                 // 生产环境构建代码
|   |-- check-version.js         // 检查 Node、NPM 等版本
|   |-- dev-client.js            // 热重载相关
|   |-- dev-server.js            // 构建本地服务器
|   |-- utils.js                 // 构建工具相关
|   |-- webpack.base.conf.js     // Webpack 基础配置
|   |-- webpack.dev.conf.js      // Webpack 开发环境配置
|   |-- webpack.prod.conf.js     // Webpack 生产环境配置
|-- config                       // 项目开发环境配置
|   |-- dev.env.js               // 开发环境变量
|   |-- index.js                 // 项目一些配置变量
|   |-- prod.env.js              // 生产环境变量
|   |-- test.env.js              // 测试环境变量
|-- node_modules                 //所需要依赖资源
|-- src                          // 源码目录
|   |-- assets                   //存放资产文件
|   |-- components               // Vue 公共组件
|   |-- router                   //存放路由 JS 文件,用于页面的跳转
|   |-- App.vue                  // 页面入口文件
|   |-- main.js                  // 程序入口文件,加载各种公共组件
|-- static                       // 静态文件,例如一些图片、JSON 数据等
```

```
|      |-- data                       // 群聊分析得到的数据用于数据可视化
|-- .babelrc                          // ES6 语法编译配置
|-- .editorconfig                     // 定义代码格式
|-- .gitignore                        // git 上传需要忽略的文件格式
|-- README.md                         // 项目说明
|-- favicon.ico
|-- index.html                        // 入口页面
|-- package.json                      // 项目基本信息
.
```

对于开发者更多操作的是 src 目录：

```
|-- src                               // 源码目录
|   |-- assets                        //存放资产文件
|   |-- components                    // Vue 公共组件
|   |-- router                        //存放路由 JS 文件,用于页面的跳转
|   |-- App.vue                       // 页面入口文件
|   |-- main.js                       // 程序入口文件,加载各种公共组件
```

main.js 文件解释：

```
/*在 main.js 中导入 Vue 对象*/
import Vue from 'vue'
/*导入 App.vue 组件,并且命名为 App*/
import App from './App'
/*导入 router 路由*/
import router from './router'
Vue.config.productionTip = false
/*所有导入成功后,创建 Vue 对象,设置被绑定的节点为'#app','#app'是 index.html 文件中的一
个 div */
new Vue({
  el: '#app',
/*将 router 设置到 Vue 对象中*/
  router,
/*声明一个组件 App,App 这个组件在一开始已经导入项目中了,但是无法直接使用,必须声明*/
  components: { App },
/*template 中定义了页面模板,即在 App 组件中的内容渲染到'#app'这个 div 中*/
  template: '<App/>'
})
```

App.vue 文件解释：

App.vue 是一个 Vue 组件,包含三部分内容：页面模板、页面脚本、页面样式,代码如下：

```html
<!-- 页面模板 -->
<template>
    <div id="app">
    <!-- 页面模板中定义了一张图片 -->
        <img src="./assets/logo.png">
        <!-- router-view 简单理解为路由占位符,用来挂载所有的路由组件 -->
        <router-view/>
    </div>
</template>
<!-- 页面脚本:页面脚本用来实现数据初始化、事件处理等 -->
<script>
    export default {
      name: 'App'
    }
</script>
<!-- 页面样式 -->
<style>
    #app {
      font-family: 'Avenir', Helvetica, Arial, sans-serif;
      -webkit-font-smoothing: antialiased;
      -moz-osx-font-smoothing: grayscale;
      text-align: center;
      color: #2c3e50;
      margin-top: 60px;
    }
</style>
```

index.js 文件解释:

```js
import Vue from 'vue'
import Router from 'vue-router'
import HelloWorld from '@/components/HelloWorld'
Vue.use(Router)
export default new Router({
/* 路由文件,path 路径,对应的组件为 HelloWorld,即在浏览器地址为/时,在 router-view 位置显
示 HelloWorld 组件 */
  routes: [
    {
      path: '/',
      name: 'HelloWorld',
      component: HelloWorld
    }
  ]
})
```

（5）在命令行输入 npm run dev 命令启动项目，如图 7-9 所示。

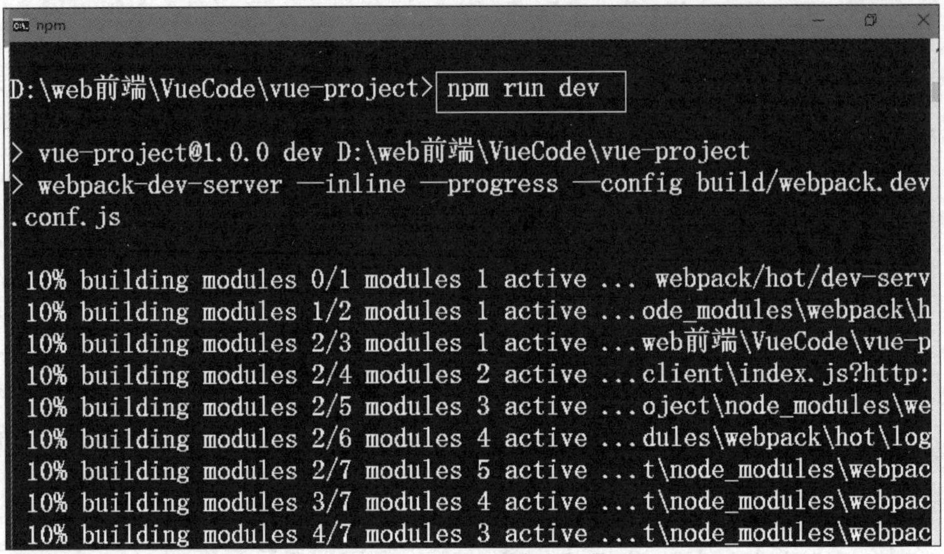

图 7-9　启动项目

运行成功后在浏览器输入 http://localhost:8080，访问项目结果如图 7-10 所示。

图 7-10　Vue 单页面效果

注意：

（1）Vue 项目是自带热部署的。

（2）如果浏览器打开之后，没有加载出页面，有可能是本地的 8080 端口被占用，此时需要修改一下配置文件 config > index.js，如图 7-11 所示。如果正常显示则忽略此操作。

（3）修改端口号是为了防止端口号被占用。

```
module.exports = {
  dev: {

    // Paths
    assetsSubDirectory: 'static',
    assetsPublicPath: './',
    proxyTable: {},

    // Various Dev Server settings
    host: 'localhost', // can be overwritten by process.env.HOST
    port: 8080, // can be overwritten by process.env.PORT, if port is in use, a free one will
    autoOpenBrowser: false,
    errorOverlay: true,
    notifyOnErrors: true,
    poll: false, // https://webpack.js.org/configuration/dev-server/#devserver-watchoptions-
```

图 7-11 index.js 配置文件

（4）把 assetsPublicPath 属性前缀修改为 './'（原本为 '/'），这是因为打包之后，外部引入 JS 和 CSS 文件时，如果路径以 '/' 开头，在本地无法找到对应的文件（在服务器上没有问题）。所以如果需要在本地打开打包后的文件，就得修改文件路径。

7.2 项目打包与发布

在控制台中输入 npm run build 命令对当前 Vue 项目进行打包，如图 7-12 所示。

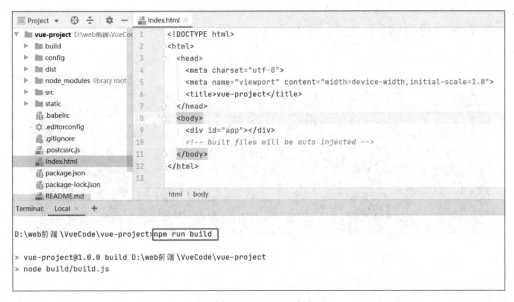

图 7-12 Vue 项目进行打包

打包完成,控制台会输出 Build complete。并且在 Vue 项目中生成一个 dist 的打包文件,如图 7-13 所示。

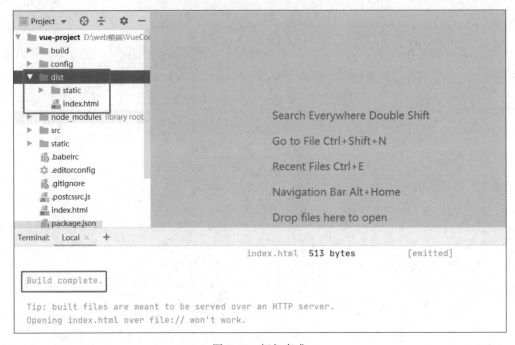

图 7-13 打包完成

7.2.1 使用静态服务器工具包发布打包

步骤 1:首先安装全局的 serve,在命令行输入命令 npm install -g serve,如图 7-14 所示。

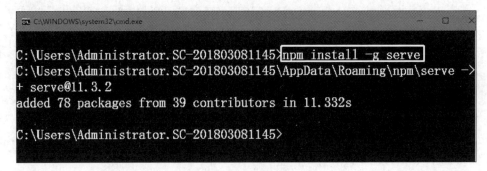

图 7-14 安装全局的 serve

步骤 2:在 WebStorm 控制台输入命令。

```
serve dist    //serve + 打包文件名
```

效果如图 7-15 所示。

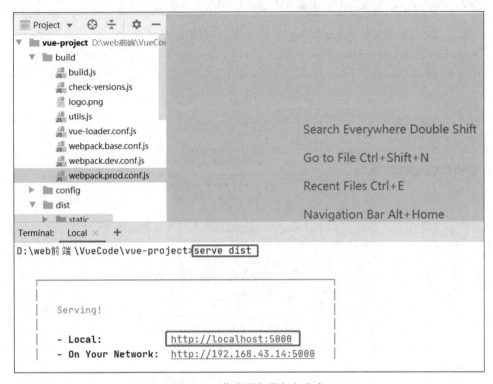

图 7-15　静态服务器打包发布

步骤 3：使用浏览器访问图 7-15 输出的地址，效果如图 7-16 所示。

图 7-16　页面显示效果

7.2.2 使用动态 Web 服务器（Tomcat）发布打包

步骤1：修改配置文件 webpack.prod.conf.js，如图 7-17 所示。

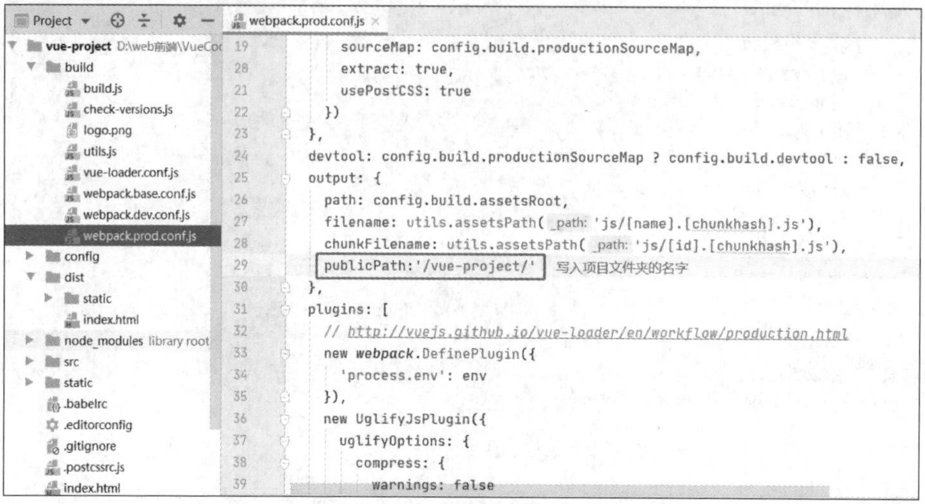

图 7-17 webpack.prod.conf.js 文件修改配置

步骤2：重新打包，如图 7-18 所示。

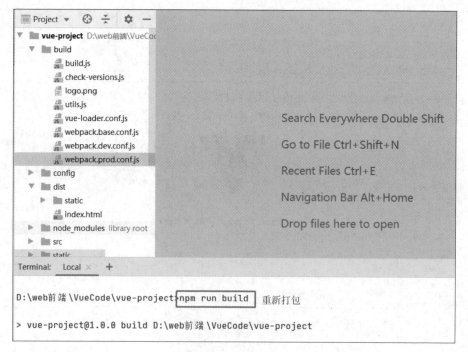

图 7-18 重新打包

步骤3：将dist文件夹复制到运行的Tomcat的webapps目录下，修改dist文件夹为项目名称（本例中为vue-project），如图7-19所示。

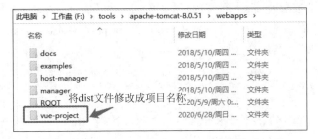

图7-19　将dist文件夹复制到运行的Tomcat的webapps目录下

步骤4：启动Tomcat，使用浏览器访问输出的地址，效果如图7-20所示。

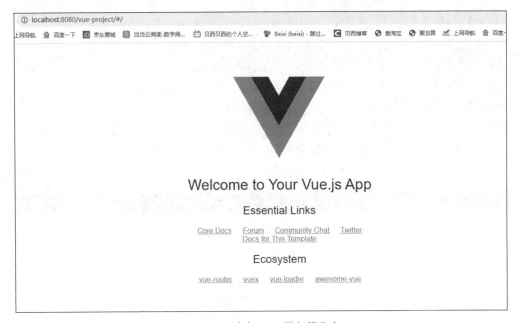

图7-20　动态Web服务器发布

7.3　Vue-devtools

在开发时经常要观察组件实例中data属性的状态，方便进行调试，但一般组件实例并不会暴露在window对象上，无法直接访问内部的data属性，若只通过debugger进行调试则效率太低。所以Vue官方推出一款Chrome插件Vue-devtools，Vue-devtools是一款基于Chrome浏览器的插件，用于调试Vue应用，这可以极大地提高调试效率。本节主要介绍Vue-devtools的安装和使用。

7.3.1　Vue-devtools 的安装

Vue-devtools 的安装步骤如下。

（1）通过 GitHub 下载 Vue-devtools 库，网址为 https://GitHub.com/vuejs/vue-devtools/tree/v5.1.1。使用 git 下载，命令如下：

```
git clone https://GitHub.com/vuejs/vue-devtools
```

（2）在 vue-devtools 目录下安装依赖包，命令如下：

```
cd vue-devtools        //进入文件目录
npm install            //如果安装太慢可以用 cnpm 代替
```

（3）编译项目文件，命令如下：

```
npm run build
```

（4）修改 manifest.json 文件，把 "persistent":false 改成 "persistent":true。一般所在路径是：自定义路径\vue-devtools-5.1.1\shells\chrome\manifest.json。

（5）使用谷歌浏览器并在地址输入栏输入 chrome://extensions/ 进入插件界面。

单击"加载已解压的扩展程序"按钮，选择 Vue-devtools > shells > chrome 放入，安装成功后如图 7-21 所示。

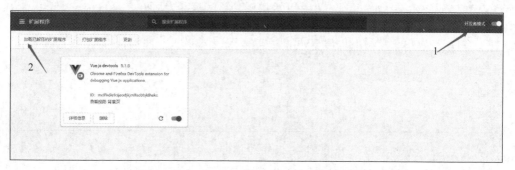

图 7-21　Vue-devtools 安装成功

7.3.2　Vue-devtools 使用

当添加完 Vue-devtools 扩展程序之后，在调试 Vue 应用的时候，打开 F12，在 Chrome 开发者工具中会看一个 Vue 栏，单击之后就可以看见当前页面 Vue 对象的一些信息，如图 7-22 所示。Vue 是通过数据驱动的，这样就能看到对应的数据了，方便进行调试。Vue-devtools 使用起来还是比较简单的，上手非常容易。

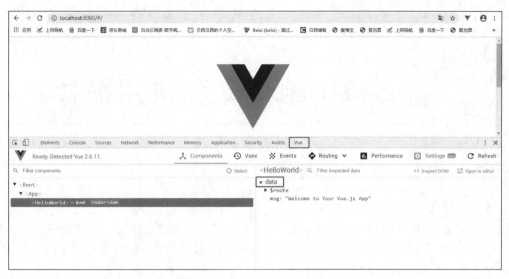

图 7-22　Vue-devtools 使用效果

第 8 章 UI 组件库和常用插件

Vue 的组件库可选择的面就广阔了许多，而对于现在前端的开发来说，我们并不需要太多地去写 Vue 框架应用本身的代码，取而代之的是使用类似组件半成品的一些组件库，例如当年首先采用响应式的 Bootstrap 或如今饿了么开源的基于移动端 Mint UI 和 PC 端 Element UI 组件库等，这些成熟、可定制的组件库让我们可以更多地把精力放在产品的构建上。

常用插件大都来自第三方开发者，是他们为 Vue 社区提供了大量的技术支持和解决方案，本章主要为 Vue 开发者讲解编写 Vue 插件的方法和步骤，通过理论与实践相结合的方式加深大家对 Vue 插件编写的认识。

8.1 Element-ui

Vue 的核心思想是组件和数据驱动，但是每一个组件都需要自己编写模板、样式、添加事件、数据等，这些工作是比较烦琐的，所以饿了么推出了基于 Vue 2.0 的组件库，它的名称为 Element-ui，此组件库提供了丰富的 PC 端组件。

ElementUI 官网为 http://element-cn.eleme.io/#/zh-CN。

Element-ui 组件库有以下四大优势。

（1）丰富的 feature：丰富的组件，自定义主题，以及国际化。
（2）文档 & demo：提供友好的文档和 demo，维护成本低，支持多语言。
（3）安装 & 引入：支持 NPM 方式和 CDN 方式安装，并支持按需引入。
（4）工程化：开发、测试、构建、部署等开发步骤持续集成。

Vue 项目中引入 Element-ui 组件库有两种方式。

1. CDN 在线方式

通过在线方式直接在页面上引入 Element-ui 的 JS 和 CSS 文件即可开始使用，代码如下：

```
<!-- 引入样式 -->
<link rel="stylesheet" href="https://unpkg.com/element-ui/lib/theme-chalk/index.css">
<!-- 引入组件库 -->
<script src="https://unpkg.com/element-ui/lib/index.js"></script>
```

2. NPM 方式安装

推荐使用 NPM 的方式安装,它能更好地和 Webpack 打包工具配合使用,后面章节中笔者也是以 NPM 的方式安装,命令如下:

```
npm install element-ui -S
```

Vue 项目中集成 Element-ui 的步骤如下。

(1) 在控制台输入命令进行安装,如图 8-1 所示。

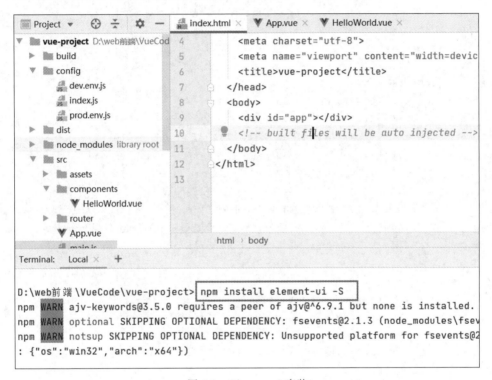

图 8-1　Element-ui 安装

(2) 查看配置文件 package.json,是否有 Element-ui 组件的版本号,如图 8-2 所示。
(3) 在 main.js 文件中完整引入 Element-ui 组件库,如图 8-3 所示。
(4) 从 Element 官网复制示例代码到 HelloWorld.vue,体验 Element-ui 组件。

在 HelloWorld.vue 中引入常用的操作按钮、下拉列表、Table 表格等组件,代码如下:

```
<template>
    <div id="app">
    <h1>{{msg}}</h1>
    <!--1.常用按钮-->
```

```html
            <el-button type = "primary">主要按钮</el-button>
            <el-button plain type = "warning">警告按钮</el-button>
<!-- 2.下拉列表 -->
<el-dropdown split-button size = "small" trigger = "click">
个人中心
<el-dropdown-menu>
<el-dropdown-item>退出系统</el-dropdown-item>
<el-dropdown-item divided>修改密码</el-dropdown-item>
            <el-dropdown-item divided>联系管理员</el-dropdown-item>
        </el-dropdown-menu>
</el-dropdown>
<br>
<!-- 3.Table 表格 -->
<el-table :data = "tableData" stripe>
 <el-table-column prop = "date" label = "日期"></el-table-column>
 <el-table-column prop = "name" label = "姓名"></el-table-column>
 <el-table-column prop = "address" label = "地址"></el-table-column>
 <el-table-column label = "操作" align = "center">
    <!--
            slot-scope:作用域插槽,可以获取表格数据
            scope:代表表格数据,可以通过 scope.row 获取表格当前行数据,scope 不是固
                定写法
    -->
    <template slot-scope = "scope">
        <el-button type = "primary" size = "mini"
            @click = "handleUpdate(scope.row)">编辑</el-button>
        <el-button type = "danger" size = "mini"
            @click = "handleDelete(scope.row)">删除</el-button>
        </template>
      </el-table-column>
    </el-table>
  </div>
</template>
<script>
export default {
  name: 'HelloWorld',
  /* 在 Vue 组件中 data 只能为函数,这是 Vue 的特性。必须由 return 返回数据,否则页面模板接收不到值 */
  data () {
    return {
```

```
      tableData:[{
        date:'2016-05-02',
        name:'王小虎',
        address:'上海市普陀区金沙江路 1518 弄'
      },{
        date:'2016-05-04',
        name:'王小虎',
        address:'上海市普陀区金沙江路 1517 弄'
      },{
        date:'2016-05-01',
        name:'王小虎',
        address:'上海市普陀区金沙江路 1519 弄'
      }]
    }
  },
  methods:{
    handleUpdate(row){
      alert(row.date);
    },
    handleDelete(row){
      alert(row.date);
    }
  }
}
</script>
```

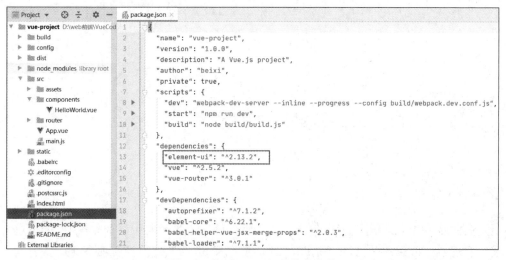

图 8-2　Element-ui 组件的版本号

```
main.js ×
1   import Vue from 'vue'
2   import App from './App'
3   import router from './router'
4   import ElementUI from 'element-ui'
5   import 'element-ui/lib/theme-chalk/index.css'
6
7   Vue.use(ElementUI)
8   Vue.config.productionTip = false
9
10  /* eslint-disable no-new */
11  new Vue({
12    el: '#app',
13    router,
14    components: { App },
15    template: '<App/>'
16  })
```

图 8-3 在 main.js 文件中完整引入 Element-ui 组件库

页面显示效果如图 8-4 所示。

图 8-4 常用组件效果展示

8.2 Vue-router

这里的路由并不是指我们平时所说的硬件路由器,这里的路由指的是 SPA(单页应用)的路径管理器。再通俗点说,Vue-router 就是 Web App 的链接路径管理系统。

Vue-router 是 Vue.js 官方的路由插件,它和 Vue.js 是深度集成的,适合用于构建单页面应用。Vue 的单页面应用是基于路由和组件的,路由用于设定访问路径,并将路径和组件映射起来。传统的页面应用使用一些超链接实现页面切换和跳转。在 Vue-router 单页面应用中,通过路径之间的切换,也就是组件的切换实现页面切换和跳转。路由模块的本质就是建立起 URL 和页面之间的映射关系。这样也有助于前后端分离,前端不用依赖后端的逻辑,只需后端提供数据接口。

那么为什么不能用 a 标签呢? 这是因为用 Vue 做的应用都是单页应用(当项目准备打包时,运行命令 npm run build 就会生成 dist 文件夹,这里面只有静态资源和一个 index.html 页面),所以我们写的<a></a>标签不起作用,必须使用 Vue-router 进行管理。

8.2.1 基本用法

回顾 7.1.3 节,在使用 Vue-cli 构建项目时选择官方推荐的自动安装 Vue-router,这样能使初学者在初次接触脚手架项目时更容易理解,能直观体验项目的全貌。自动安装 Vue-router 项目后 src 目录下会多出一个 router 文件夹,并且 App.vue 的内容多出一对<router-view></router-view>标签,router-view 标签主要将路由路径所指定的组件渲染到页面中,此时可以将所有的路由路径写在 router 文件夹里的 index.js 文件里,实现在 URL 上输入不同的路径从而渲染不同组件的功能。

现在使用 Webpack 模板构建项目(如:vue-project2),如图 8-5 所示,选择不安装 Vue-router。

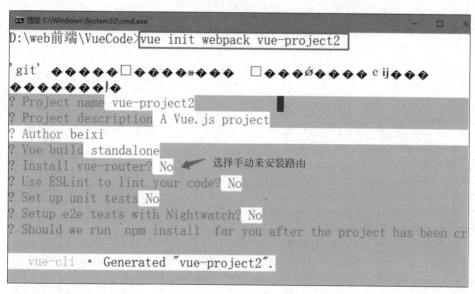

图 8-5 使用 Webpack 模板构建项目 vue-project2

通过命令 npm install 安装完依赖后,项目目录结构如图 8-6 所示。
通过命令 npm install 安装 Vue-router,命令如下:

```
npm install -- save vue - router    //进入项目根目录后安装路由依赖
```

图 8-6 项目目录结构

通过上述命令安装完路由依赖后,使用步骤如下:

(1) 在 src 目录下建立 router/index.js 文件,此文件专门用于管理路由核心文件,使用 Vue.use()加载路由插件,代码如下:

```
//引入 Vue 框架
import Vue from 'vue'
//引入 Vue-router 路由依赖
import Router from 'vue-router'
//引入页面组件,命名为 HelloWorld。@代表绝对路径
import HelloWorld from '@/components/HelloWorld'
//Vue 全局使用 Router
Vue.use(Router)
//定义路由配置,注意 export 导出,只要导出了别的文件才能 import 导入
export default new Router({
    routes: [                    //配置路由,这里是一个数组
        {                        //每一个链接都是一个对象
            //链接路径,浏览器路径是 localhost:8080/时,链接到 HelloWorld.vue 组件
            path: '/',
            name: 'HelloWorld',  //路由名称
            component: HelloWorld//对应的组件模板
        }
    ]
})
```

（2）在系统入口文件 main.js 中导入 router，代码如下：

```
import Vue from 'vue'
import App from './App'
//引入路由,自动会寻找 index.js
import router from './router'
//关闭生产模式下给出的提示
Vue.config.productionTip = false
/* eslint-disable no-new */
new Vue({
  el: '#app',
  router,         //注入框架中。此处是简写,key 和 value 一致,等价于 router:router
  components: { App },
  template: '<App/>'
})
```

（3）用 router-view 占位符定义路由出口，路由匹配到的组件内容将渲染到这里，具体在 App.vue 组件中的用法代码如下：

```
<template>
    <div id="app">
        <img src="./assets/logo.png">
        <!-- 路由占位符:将路由路径所指定的组件渲染到页面中 -->
        <router-view/>
    </div>
</template>
<script>
    /* 注释掉 */
    //import HelloWorld from './components/HelloWorld'
    export default {
      name: 'App'
    }
</script>
<style>
#app {
  font-family: 'Avenir', Helvetica, Arial, sans-serif;
  -webkit-font-smoothing: antialiased;
  -moz-osx-font-smoothing: grayscale;
  text-align: center;
  color: #2c3e50;
  margin-top: 60px;
}
</style>
```

（4）简化 HelloWorld.vue 组件内容，代码如下：

```
<template>
    <div id="hello">
        <h1>HelloWorld.vue组件</h1>
    </div>
</template>
<script>
export default {
  name: 'HelloWorld',
  data () {       //在组件中必须为 data 函数，这是 Vue 的特性
      return {
           msg: 'Welcome to Your Vue.js App'
      }
  }
}
</script>
```

（5）运行命令 pm run dev 启动服务，访问 http://localhost:8080/，页面显示效果如图 8-7 所示。

图 8-7　效果运行效果

Vue 提供了一个属性 mode:"history"，可以去掉地址栏中的 #，代码如下：

```
//index.js      部分代码省略
export default new Router({
   mode:'history',         //去掉路由地址栏中的 #
   routes: [               //配置路由，这里是一个数组
     {                     //每一个链接都是一个对象
         //跳转路径，浏览器路径是 localhost:8080/时,跳转到 HelloWorld.vue 组件
       path: '/',
       name: 'HelloWorld',  //路由名称,
       component: HelloWorld //对应的组件模板
     }
   ]
})
```

设置 mode 为 history 会开启 HTML5 的 History 路由模式,通过"/"设置路径。如果不设置 mode,就会使用"#"设置路径。

每个页面对应一个组件,也就是对应一个 .vue 文件。在 components 目录下创建 Hi.vue 文件,代码如下:

```
<template>
    <div id="hi">
        <h1>Hi.vue 组件</h1>
    </div>
</template>
<script>
    export default {
        name: "Hi"
    }
</script>
```

再回到 index.js 里,完成路由的配置,完整代码如下:

```
//引入 Hi 组件
import Hi from '../components/Hi'
//Vue 全局使用 Router
Vue.use(Router)
//定义路由配置,注意 export 导出,只要导出了别的文件才能 import 导入
export default new Router({
    mode:'history',          //去掉路由地址栏中的#
    routes: [                //配置路由,这里是一个数组
//...
        {
            path: '/hi',
            name: 'Hi',
            component: Hi
        }
    ]
})
```

启动服务,在浏览器中访问 http://localhost:8080/ 和 http://localhost:8080/hi 就可以分别访问 HelloWorld.vue 和 Hi.vue 组件页面了。

ES6 语法提示:

ES6 新增了 let 和 const 命令声明变量,代替了 var。它们的用法类似于 var,但是所声明的变量只在代码块内有效,代码如下:

```
{
  let a = 123;
```

```
    var b = 234;
}
console.log(a);         //报错 a is not defined
console.log(b);         //234
```

let 和 const 的主要区别是,const 声明的是一个只读常量。一旦声明,常量的值就不能改变。

8.2.2 跳转

在 components 目录下创建 Home.vue(首页)和 News.vue(新闻页)两个文件,代码如下:

```
//Home.vue
<template>
    <div id="Home">
        <h1>我是首页</h1>
    </div>
</template>
<script>
    export default {
        name:"Home"
    }
</script>

//News.vue
<template>
    <div id="News">
        <h3>{{title}}</h3>
        <ul class="ulnews">
            <li v-for="(data,index) in newslist">{{data}}</li>
        </ul>
    </div>
</template>
<script>
    export default {
        name:"News",
        data(){
            return{
                title:'新闻栏目',
                newslist:[
                    '新闻1',
                    '新闻2',
                    '新闻3'
```

```
                ]
            }
        }
    }
</script>
<style scoped>
    .ulnews li{
        display: block;
    }
</style>
```

接着在 index.js 中完成路由配置,代码如下:

```
//router/index.js 部分代码省略
import Home from '@/components/Home' //导入 Home.vue
import News from '@/components/News' //导入 News.vue
export default new Router({
    mode:'history',
    routes: [           //配置路由匹配列表
//...
    {
        path: '/home',
        name: 'Home',
        component: Home
    },
    {
        path: '/news',
        name: 'News',
        component: News
    }
    ]
})
```

这样我们便可以在地址栏输入不同的路径访问对应的页面,但页面上一般需要有导航链接,我们只要单击此链接就可以实现页面内容的变化。在 components 目录下创建 Nav.vue 导航页面,代码如下:

```
//Nav.vue
<template>
    <div id="box">
    <ul>
        <li>首页</li>
        <li>新闻</li>
    </ul>
```

```
        </div>
    </template>
    <script>
        export default {
            name: "nav"
        }
    </script>
    <style scoped>
      *{
        padding: 0;
        margin: 0;
      }
      ul{
        list-style: none;
        overflow: hidden;
      }
      #box{
        width: 600px;
        margin: 100px auto;


      }
      #box ul{
        padding: 0 100px;
        background-color: #2dc3e8;
      }
      #box ul li {
        display: block;
        width: 100px;
        height: 50px;
        background-color: #2dc3e8;
        color: #fff;
        float: left;
        line-height:50px;
        text-align: center;
      }
      #box ul li:hover{
        background-color: #00b3ee;
      }
    </style>
```

导航页面构建完成后,在实例 App.vue 组件中导入,这样我们在直接访问 http://localhost:8080/路径时就会显示导航栏,代码如下:

```
//App.vue
<template>
```

```
    <div id = "app">
        <Nav></Nav>
        <router-view/><!-- 路由占位符:将路由路径所指定的组件渲染到页面中 -->
    </div>
</template>
<script>
import Nav from '@/components/Nav'
export default {
  name: 'App',
  components:{
    Nav
  }
}
</script>
```

页面显示效果如图 8-8 所示。

图 8-8 导航栏效果

下面介绍 Vue.js 通过路由跳转页面方式。

1. router-link 标签跳转

使用内置的< router-link >标签跳转,它会被渲染成< a >标签,使用方式如下:

```
//Nav.vue
<template>
    <div id = "box">
        <ul>
            <li><router-link to = "/home">首页</router-link></li>
            <li><router-link to = "/news">新闻</router-link></li>
        </ul>
    </div>
</template>
```

<router-link>标签中的属性介绍如下。

- to

表示目标路由的链接。当被单击后,内部会立刻把 to 的值传到 router.push(),所以这个值可以是一个字符串,也可以是描述目标位置的对象。当然也可以用 v-bind 动态设置链接路径,代码如下:

```
<li><router-link to="/home">首页</router-link></li>
<!-- v-bind 动态设置 -->
<li><router-link v-bind:to="home">首页</router-link></li>
<!-- 同上 -->
<li><router-link :to="{ path: 'home' }">首页</router-link></li>
```

- replace

使用 replace 属性导航后不会留下 history 记录,所以导航后不能用后退键返回上一个页面,如<router-link to="/home" replace>。

- tag

有时需要使<router-link>渲染成某种标签,例如<li>。我们可以使用 tag 属性指定标签,同样它可以监听单击,从而触发导航,代码如下:

```
<router-link to="/foo" tag="li">foo</router-link>
```

渲染结果为:<li>foo</li>。

- active-class

当<router-link>对应的路由匹配成功时,会自动给当前元素设置一个名为 router-link-active 的 class 值,如图 8-9 所示。

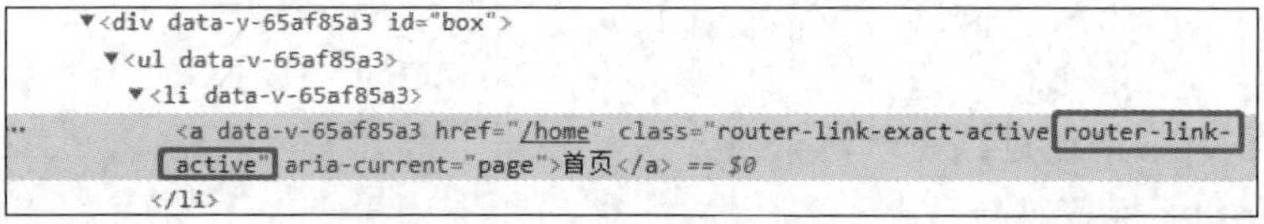

```
▼<div data-v-65af85a3 id="box">
  ▼<ul data-v-65af85a3>
    ▼<li data-v-65af85a3>
      ... <a data-v-65af85a3 href="/home" class="router-link-exact-active router-link-active" aria-current="page">首页</a> == $0
      </li>
```

图 8-9 自动设置名为 router-link-active 的 class 值

设置 router-link-active 的 CSS 样式。在单击导航栏时,可以使用该功能高亮显示当前页面对应的导航菜单项。

当我们访问 http://localhost:8080/路径时,希望在页面中显示首页的内容,而不是 HelloWorld.vue 组件的内容。这时我们可以使用路由的重新定向 redirect 属性。我们在路由配置文件中设置 redirect 参数即可,代码如下:

```
//router/index.js 部分代码省略
export default new Router({
```

```
routes: [
  {
    path: '/',
    redirect:'/home',
    name: 'HelloWorld',
    component: HelloWorld
  }
//...
  ]
})
```

2. JS 代码内部跳转

在实际项目中,很多时候通过在 JS 代码内部进行导航的跳转,使用方式如下:

```
this.$router.push('/xxx')
```

具体的用法及步骤如下:

(1) 在页面中添加单击事件,代码如下:

```
<button @click = "goHome">首页</button>
```

(2) 在<script>模块里加入 goHome 方法,并用 this.$router.push('/')导航到首页,代码如下:

```
//部分代码省略
export default {
    name: 'app',
    methods: {
        goHome(){
            this.$router.push('/home');
        }
    }
}
```

其他常用方法:

```
//后退一步记录,等同于 history.back()
this.$router.go(-1)
//在浏览器记录中前进一步,等同于 history.forward()
this.$router.go(1)
```

8.2.3 路由嵌套

子路由也叫路由嵌套,一般应用都会出现二级导航这种情况,这时就得使用路由嵌套这种写法，采用在 children 后跟路由数组来实现，数组的配置和其他配置路由基本相同，需要配置 path 和 component，然后在相应部分添加<router-view/>展现子页面信息，相当于嵌入 iframe。具体如下面的示例：

（1）Src/components/Home.vue（父页面），代码如下：

```
<template>
    <div id="Home">
        <h1>我是首页：</h1>
        <ul>
            <!-- 路径要写完整路径：父路径 + 子路径 -->
            <span><router-link to = "/home/one">子页面 1</router-link></span>
            <span><router-link to = "/home/two">子页面 2</router-link></span>
        </ul>
        <!-- 子页面展示部分 -->
        <router-view/>
    </div>
</template>
<script>
    export default {
        name: "Home"
    }
</script>
<style scoped>
  .router-link-active{
    color: red;
  }
</style>
```

（2）src/pages/One.vue（子页面 1），代码如下：

```
<template>
    <div id = "one">
        <h3>{{msg}}</h3>
</div>
</template>
<script>
    export default {
        name: "One",
      data(){
        return {
           msg:'这是第一个子页面'
        }
      }
    }
</script>
```

(3) src/pages/Two.vue(子页面2),代码如下:

```html
<template>
    <div id="two"><h3>{{msg}}</h3></div>
</template>
<script>
    export default {
        name: "Two",
        data(){
          return {
             msg:'这是第二个子页面'
          }
        }
    }
</script>
```

(4) src/router/index.js(路由配置),代码如下:

```js
//部分代码省略
/* 导入页面1和页面2 */
import One from '@/pages/One'
import Two from '@/pages/Two'
export default new Router({
    routes: [
//...
    {
       path: '/home',
       name: 'Home',
       component: Home,
       children:[         //嵌套子路由
         {
           path:'one',    //子页面1
           component:One
         },
         {
           path:'two',    //子页面2
           component:Two
         },
       ] }
    ]
})
```

(5) 页面显示效果如图8-10所示。

注意:子页面显示的位置在其父级页面中,所以一定要在父级页面中添加<router-view/>标签。

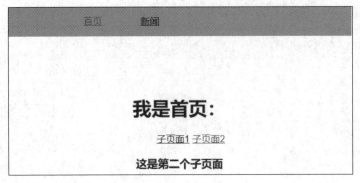

图 8-10　路由嵌套效果图

8.2.4　路由参数传递

路由传递参数有 3 种方式。

1．通过< router-link >标签中的 to 传参

基本语法：

```
< router - link :to = "{name:xxx,
params:{key:value}}"> valueString </router - link >
```

上面代码中 to 前面带冒号，值为对象形式的字符串。

（1）name：对应的是路由配置文件中所使用的 name 值，叫作路由名称。

（2）params：要传的参数，此参数以对象形式出现，在对象里可以传递多个值。

具体实例如下：

（1）在 src/components/Home.vue 里面的导航中添加如下代码。

```
< router - link :to = "{name: 'One', params:{username:'beixi'}}">
子页面 1 </router - link >
```

（2）在 src/router/indes.js 中添加如下代码：

```
{
  path:'one',       //子页面 1
  name:'One',       //name 属性不能少
  component:One
}
```

（3）在 src/pages/One.vue 里面接收参数，代码如下：

```
< h2 >{{ $ route.params.username}}</h2>
```

2. 在 URL 中传递参数

（1）在路由中以冒号形式传递，在 src/router/index.js 中添加如下代码：

```
{
  path:'two/:id/:name', //子页面 2
  component:Two
}
```

（2）接收参数，在 src/pages/Two.vue 中添加如下代码：

```
<p>ID: {{ $route.params.id}}</p>
<p>名称: {{ $route.params.name}}</p>
```

（3）路由跳转，在 src/components/Home.vue 中添加如下代码：

```
<router-link to = "/home/two/666/贝西奇谈">子页面 2</router-link>
```

3. 编程式导航 params 传递参数

（1）在 src/router/index.js 页面添加如下代码：

```
{
  path:'three/:id/:name',      //子页面 3
  name: 'three',
  component:Three
}
```

（2）在 src/pages/Three.vue 页面添加如下代码：

```
<p>ID: {{ $route.params.id}}</p>
<p>名称: {{ $route.params.name}}</p>
```

（3）在 src/components/Home.vue 中添加如下代码：

```
//template 部分代码省略
<button @click = "toThreePage">页面 3</button>
//script
methods: {
    toThreePage() {
        this.$router.push({name: 'three', params: {id: 1, name:'beixi'}})
    }
}
```

代码说明：

（1）动态路由使用 params 传递参数，在 this.$router.push() 方法中 path 不能和 params

一起使用,否则 params 将无效。需要用 name 来指定页面。

（2）以上方式传递的参数不会显示到浏览器的地址栏中,如果刷新一次页面,就获取不到参数了,改进方式如下：

```
{
    path:'/home/three/:id/:name',      //子页面 3
    name: 'three',
    component:Three
}
```

路由中 router 和 route 的区别如下：

（1）router 是 VueRouter 的一个对象,通过 Vue.use(VueRouter)和 VueRouter 构造函数得到一个 router 的实例对象,这个对象是一个全局的对象,它包含了所有的路由并且也包含了许多关键的对象和属性,代码如下：

```
$ router.push({path:'home'});          //切换路由,有 history 记录
$ router.replace({path:'home'});       //替换路由,没有历史记录
```

（2）route 是一个跳转的路由对象,每一个路由都会有一个 route 对象,它是一个局部的对象,可以获取对应的 name、path、params、query 等。

$ route.path、$ route.params、$ route.name、$ route.query 这几个属性很容易理解,主要用于接收路由中传递的参数。

8.3 Axios

Axios 是一个基于 Promise 的 HTTP 库,它是一个简洁、易用且高效的代码封装库。通俗地讲,它是当前比较流行的一种 Ajax 框架,可以使用它发起 HTTP 请求接口功能,它是基于 Promise 的,相比较 Ajax 的回调函数能够更好地管理异步操作。

Axios 的特点：

（1）从浏览器中创建 XMLHttpRequests。

（2）从 Node.js 创建 HTTP 请求。

（3）支持 Promise API。

（4）拦截请求和响应。

（5）转换请求数据和响应数据。

（6）取消请求。

（7）自动转换 JSON 数据。

（8）客户端支持防御 XSRF。

8.3.1 基本使用

首先使用 NPM 安装 Axios 的依赖，命令如下：

```
npm install axios
```

如果要全局使用 Axios 就需要在 main.js 中设置，然后在组件中通过 this 调用，代码如下：

```
import axios from 'axios'
Vue.prototype.$axios = axios;        //加载到原型上
```

Axios 提供了很多请求方式，例如在组件中直接发起 GET 或 POST 请求：

```
//为给定 ID 的 user 创建 GET 请求
    this.$axios.get('/user?ID=12345').then(res=>{console.log('数据是:', res);})
//也可以把参数提取出来
    this.$axios.get('/user',{
        params: {
            ID: 12345
        }
    })
    .then(res => {          //响应数据
        console.log('数据是:', res);
    })
    .catch((e) => {         //请求失败
        console.log('获取数据失败');
    });
    //POST 请求
    this.$axios.post('/user',{
        firstName: 'Fred',
        lastName: 'Flintstone'
    })
    .then(function(res){
        console.log(res);
    })
    .catch(function(err){
        console.log(err);
    });
```

分别写两个请求函数，利用 Axios 的 all 方法接收一个由每个请求函数组成的数组，可以一次性发送多个请求，如果全部请求成功，在 axios.spread 方法中接收一个回调函数，该函数的参数就是每个请求返回的结果，代码如下：

```
function getUserAccount(){
  return axios.get('/user/12345');
}
function getUserPermissions(){
  return axios.get('/user/12345/permissions');
}
this.$axios.all([getUserAccount(),getUserPermissions()])
  .then(axios.spread(function(res1,res2){
     //当这两个请求都完成的时候会触发这个函数,两个参数分别代表返回的结果
}))
```

以上通过 Axios 直接发起对应的请求其实是 Axios 为了方便起见给不同的请求提供别名方法。我们完全可以通过调用 Axios 的 API,传递一个配置对象来发起请求。

发送 POST 请求,参数写在 data 属性中,代码如下:

```
axios({
  url: 'http://rap2api.taobao.org/app/mock/121145/post',
  method: 'post',
  data: {
    name: '小月'
  }}).then(res => {
  console.log('请求结果: ', res);});
```

发送 GET 请求,默认就是 GET 请求,直接在第一个参数处写路径,在第二个参数处写配置对象,参数通过 params 属性设置,代码如下:

```
axios('http://rap2api.taobao.org/app/mock/121145/get', {
  params: {
    name: '小月'
  }}).then(res => {
  console.log('请求结果: ', res);});
```

Axios 为所有请求方式都提供了别名:

```
axios.request(config)
axios.get(url, [config])
axios.delete(url, [config])
axios.head(url, [config])
axios.options(url, [config])
axios.post(url, [data], [config])
axios.put(url, [data], [config])
axios.patch(url, [data], [config])
```

注意:在使用别名方法时,url、method、data 这些属性都不必在配置中指定。

8.3.2　json-server 的安装及使用

json-server 是一个 Node 模块,运行 Express 服务器,我们可以指定一个 JSON 文件作为 API 的数据源。简单理解为在本地创建数据接口,使用 Axios 访问数据,使用步骤如下。

(1) 首先输入命令 cmd 进入终端,在根目录下全局安装 json-server,命令如下:

```
npm install -g json-server
```

(2) 在任意盘符中创建一个文件夹用于存放 JSON 数据文件,终端切换到该文件目录下,执行初始化命令(一直按回车键即可):

```
npm init
```

(3) 在初始化的项目中安装 json-server,执行如下命令:

```
npm install json-server --save
```

(4) 此时我们在项目文件夹下就可看到一个 package.json 文件,然后在当前目录下新建一个 db.json 文件,在本文件下编写自己的 JSON 数据(例如我的数据):

```
{
"users":[
    {
        "name":"beixi",
        "phone":"1553681223*",
        "email":"635498720@qq.com",
        "id":1,
        "age":18,
        "companyId":1
    },{
        "name":"jzj",
        "phone":"1553681223*",
        "email":"jzj@email.com",
        "id":2,
        "age":34,
        "companyId":2
    },{
        "name":"beixiqitan",
        "phone":"1553681223*",
        "email":"beixiqitan@email.com",
        "id":3,
        "age":43,
        "companyId":3
```

```
        }
    ],"companies":[
            {
                "id":1,
                "name":"Alibaba",
                "description":"Alibaba is good"
            },{
                "id":2,
                "name":"Miciosoft",
                "description":"Microsoft is good"
            },{
                "id":3,
                "name":"Google",
                "description":"Google is good"
            }
    ]
}
```

（5）修改 package.json 数据，设置快捷启动 json-server 命令如下：

```
{
    "name": "jsondemo",
    "version": "1.0.0",
    "description": "",
    "main": "index.js",
    "scripts": {
            "json:server": "json-server --watch db.json"
    },
"author": "",
"license": "ISC",
"dependencies": {
            "json-server": "^0.16.1"
    }
}
```

（6）运行 json-server，命令如下：

```
npm run json:server
```

接着我们利用 Axios 访问 json-server 服务器中的数据，对数据列表进行增、删、查操作。完整实例代码如下：

```
//src/components/jsonDemo.vue
<template>
```

```html
<div id="jsonDemo" v-cloak>
    <div class="add">
    用户信息:
    <input type="text" v-model="name">
    <input type="button" value="添加" @click=addItem()>
    </div>
    <div>
        <table class="tb">
            <tr>
                <th>编号</th>
                <th>用户名称</th>
                <th>操作</th>
            </tr>
            <tr v-for="(v,i) in list" :key="i">
                <td>{{i+1}}</td>
                <td>{{v.name}}</td>
                <td>
                    <a href="#" @click.prevent="deleItem2(v.id)">删除</a>
                </td>
            </tr>
            <!-- v-if="条件表达式" -->
            <tr v-if="list.length === 0">
                <td colspan="4">没有数据</td>
            </tr>
        </table>
    </div>
</div>
</template>
<script>
    export default {
        name: "jsonDemo",
        data(){
            return{
                name:'',
                list:[]
            }
        },
        created(){
            this.$axios.get("http://localhost:3000/users").then(res => {
                this.list = res.data
                console.log(res)
            })
            .catch(error => {
                console.log(error);
            })
```

```js
        },
        methods:{
            getData() {            //获取数据
            this.$axios.get('http://localhost:3000/users')
              .then((res) => {
                const { status, data } = res;
                if (status === 200) {
                  this.list = data;
                }
              })
              .catch((err) => {
              })
            },
            deleItem2(ID) {         //删除
            console.log(ID)
             if (confirm("确定要删除吗?")) {
               // this.list.splice(index, 1);
               this.$axios.delete('http://localhost:3000/users/' + ID)
                 .then((res) => {
                   console.log(res);
                   this.getData()
                 })
             }
            },
            addItem() {            //增加
            this.$axios.post('http://localhost:3000/users', {
                name: this.name
              })
              .then((res) => {
                const { status } = res;
                if (status === 201) {
                  this.getData()
                }
              })
            this.list.unshift({
                name: this.name,
              });
            this.name = "";
            }
          }
        }
</script>
<style scoped>
  #app {
    width: 600px;
```

```css
      margin: 10px auto;
    }
    .tb {
      border-collapse: collapse;
      width: 100%;
    }
    .tb th {
      background-color: #0094ff;
      color: white;
    }
    .tb td,
    .tb th {
      padding: 5px;
      border: 1px solid black;
      text-align: center;
    }

    .add {
      padding: 5px;
      border: 1px solid black;
      margin-bottom: 10px;
    }
</style>
```

8.3.3　跨域处理

跨域是指浏览器不能执行其他网站的脚本。它是由浏览器的同源策略造成的,是浏览器对 JavaScript 实施的安全限制。

同源策略：是指协议、域名、端口都要相同,其中有一个不同便会产生跨域。

如我们调用百度音乐的接口,以此获取音乐数据列表,这必然会出现跨域问题。百度音乐的接口完整网址：http://tingapi.ting.baidu.com/v1/restserver/ting? method=baidu.ting.billboard.billList&type=1&size=10&offset=0,如图 8-11 所示。

在 Vue 中使用本地代理的方式进行跨域处理。首先在配置文件 config/index.js 里设置代理,在 proxyTable 中添加如下代码：

```
//config/index.js 部分代码省略
module.exports = {
  dev: {
    proxyTable: {
      '/api': {
      //目标路径,别忘了添加 http 和端口号
        target: 'http://tingapi.ting.baidu.com',
```

```
        changeOrigin: true,      //是否跨域,true 为跨域
        pathRewrite: {
          '^SymbolYCp/api': ''   //重写路径
        }
      }
    }
}});
```

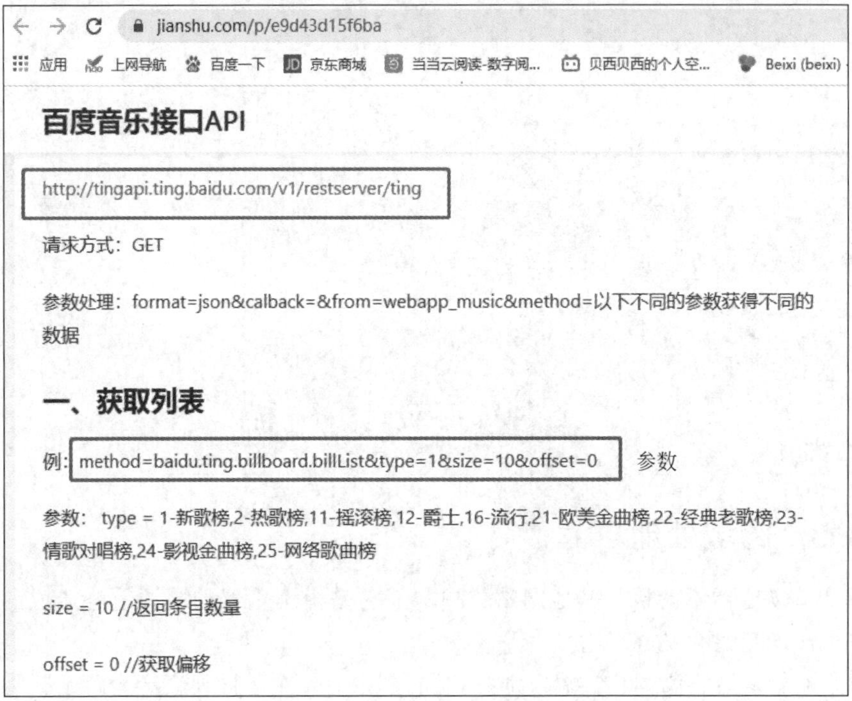

图 8-11 百度音乐列表网址

在 main.js 中,配置访问的 URL 前缀,这样每次发送请求时都会在 URL 前自动添加/api 的路径,代码如下:

```
//全局设置基本路径/api 代表 http://tingapi.ting.baidu.com 路径
axios.defaults.baseURL = '/api'
```

在 src/components/baidu.vue 建立组件页面,发送 URL 请求,代码如下:

```
<template lang = "html">
  <div class = "">
  <h3>百度音乐</h3>
  <ul>
```

```html
        <!-- 音乐集合    获取音乐标题 -->
    <li v-for="(item,index) in music.song_list">{{ item.album_title}}</li>
  </ul>
  </div>
</template>
<script>
export default {
  name:"baidu",
  data(){
    return{
      music:{
        song_list:[]
      },
    }
  },
  created(){
    //网络请求
            var url = "/v1/restserver/ting?method = baidu.ting.billboard.billList&type = 1&size = 10&offset = 0";
    this. $ axios.get(url)
    .then(res => {
      this.music = res.data
      console.log(res)
    })
    .catch(error => {
      console.log(error);
    })
  },
}
</script>
<style>
</style>
```

在 src/router/index.js 中配置路由,代码如下:

```
//src/router/index.js  部分代码省略
import baidu from '@/components/baidu'
export default new Router({
    routes: [
    {
      path: '/baidu',
      name: 'baidu',
      component: baidu
    }
  ]
})
```

重启服务,运行结果如图 8-12 所示。

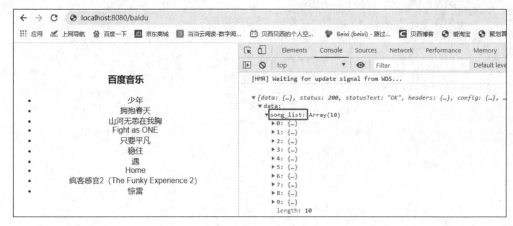

图 8-12　页面显示效果

8.3.4　Vue 中 Axios 的封装

1. Axios 的封装

在项目中新建 api/index.js 文件,用以封装配置 Axios,代码如下:

```
let http = axios.create({
  baseURL: 'http://localhost:8080/',
  withCredentials: true,
  headers: {
    'Content-Type': 'application/x-www-form-urlencoded;charset=utf-8'
  },
  transformRequest: [function (data) {
    let newData = '';
    for (let k in data) {
      if (data.hasOwnProperty(k) === true) {
        newData += encodeURIComponent(k) + '=' + encodeURIComponent(data[k]) + '&';
      }
    }
    return newData;
  }]
});
function apiAxios(method, url, params, response) {
  http({
    method: method,
    url: url,
    data: method === 'POST' || method === 'PUT' ? params : null,
    params: method === 'GET' || method === 'DELETE' ? params : null,
```

```
    }).then(function (res) {
      response(res);
    }).catch(function (err) {
      response(err);
    })
  }
  export default {
    get: function (url, params, response) {
      return apiAxios('GET', url, params, response)
    },
    post: function (url, params, response) {
      return apiAxios('POST', url, params, response)
    },
    put: function (url, params, response) {
      return apiAxios('PUT', url, params, response)
    },
    delete: function (url, params, response) {
      return apiAxios('DELETE', url, params, response)
    }
  }
```

这里配置了 POST、GET、PUT、DELETE 方法,并且自动将 JSON 格式数据转为 URL 拼接的方式。同时配置了跨域,如果不需要,则将 withCredentials 设置为 false。设置默认前缀地址为 http://localhost:8080/,这样调用的时候只需写目标后缀路径。

注意:PUT 请求默认会发送两次请求,第一次预检查请求不含参数,所以后端不能对 PUT 请求地址做参数限制。

2. 使用

首先在 main.js 中引入方法,代码如下:

```
//main.js 部分代码省略
import Api from './api/index.js';
Vue.prototype.$api = Api;
```

然后在需要的地方调用即可,代码如下:

```
this.$api.post('user/login.do(地址)', {
"参数名": "参数值"}, response => {
    if (response.status >= 200 && response.status < 300) {
        console.log(response.data);      \\请求成功,response 为成功信息参数
    } else {
        console.log(response.message);   \\请求失败,response 为失败信息
    }});
```

8.4 Vuex

Vuex 是一个专为 Vue.js 应用程序开发的状态管理工具。它采用集中式存储并管理应用的所有组件的状态,并以相应的规则保证状态以一种可预测的方式发生变化。

在学习 Vue.js 时,大家一定知道在 Vue 中各个组件之间传值的痛苦,在 Vue 中我们可以使用 Vuex 保存需要管理的状态值,值一旦被修改,所有引用该值的地方都会自动更新。

8.4.1 初识 Vuex

Vuex 适用于在 Vue 项目开发时使用的状态管理工具。试想一下,如果在一个项目开发中频繁地使用组件传参的方式同步 data 中的值,一旦项目变得很庞大,管理和维护这些值将是相当棘手的工作。为此,Vue 为这些被多个组件频繁使用的值提供了一个统一的管理工具——Vuex。在具有 Vuex 的 Vue 项目中,我们只需把这些值定义在 Vuex 中,即可在整个 Vue 项目的组件中使用。

Vuex 也被集成到 Vue 的官方调试工具 devtools extension 中了,提供了诸如零配置的 time-travel 调试、状态快照导入及导出等高级调试功能。

状态管理:简单理解就是统一管理和维护各个 Vue 组件的可变化状态(可以理解成 Vue 组件里的某些 data)。

让我们从一个简单的 Vue 计数应用开始学习,代码如下:

```
//src/components/count.vue
<template>
    <div>
        <!-- view -->
        <div>{{ count }}</div>
        <button @click="increment">increment</button>
    </div>
</template>
<script>
export default {
    //state
    data () {
        return {
            count: 0
        }
    },
    //actions
    methods: {
        increment () {
```

```
            this.count++
        }
    }}
</script>
```

这个状态管理应用包含以下几个部分。

（1）State：驱动应用的数据源，通常用在 data。

（2）View：以声明方式将 state 映射到视图，通常用在 template。

（3）Actions：响应在 View 上的用户输入导致的状态变化，即通过方法对数据进行操作，通常用 methods。

以下是一个表示"单向数据流"理念的极简示意，如图 8-13 所示。

这里的数据 count 和方法 increment 只有在 count.vue 组件里可以访问和使用，其他的组件无法读取和修改 count。但是，当我们的应用遇到多个组件共享状态时，单向数据流的简洁性很容易被破坏：

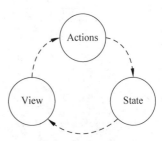

图 8-13　单向数据流

（1）多个视图依赖于同一状态。传参的方法对于多层嵌套的组件将会非常烦琐，并且对于兄弟组件间的状态传递无能为力。

（2）来自不同视图的行为需要变更同一状态。我们经常会采用父子组件直接引用或者通过事件变更和同步状态的多份拷贝。以上的这些模式非常脆弱，通常会导致无法维护这些代码。

因此，我们为什么不把组件的共享状态抽取出来，以一个全局单例模式管理呢？在这种模式下，我们的组件树构成了一个巨大的"视图"，不管在树的哪个位置，任何组件都能获取状态或者触发行为。

另外，通过定义和隔离状态管理中的各种概念并强制遵守一定的规则，我们的代码将会变得更结构化且易维护。

这就是 Vuex 背后的基本思想，借鉴了 Flux、Redux 和 The Elm Architecture。与其他模式不同的是，Vuex 是专门为 Vue.js 设计的状态管理库，以利用 Vue.js 的细粒度数据响应机制进行高效的状态更新，如图 8-14 所示。

示意图说明：

（1）Vue Components：Vue 组件。在 HTML 页面上，负责接收用户操作等交互行为，执行 dispatch 方法触发对应的 action 进行回应。

（2）Dispatch：操作行为触发方法，是唯一能执行 action 的方法。

（3）Actions：操作行为处理模块。负责处理 Vue Components 接收的所有交互行为。包含同步/异步操作，支持多个同名方法，按照注册的顺序依次触发。向后台 API 请求的操

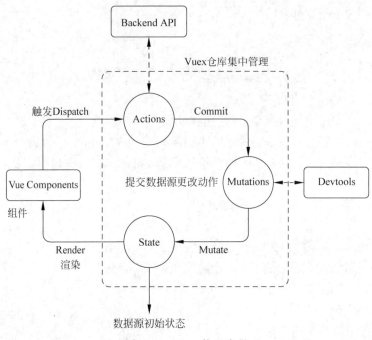

图 8-14 Vuex 核心流程

作就在这个模块中进行，包括触发其他 action 及提交 mutation 的操作。该模块提供了 Promise 的封装，以支持 action 的链式触发。

（4）Commit：状态改变提交操作方法。对 mutation 进行提交，是唯一能执行 mutation 的方法。

（5）Mutations：状态改变操作方法。是 Vuex 修改 state 的唯一推荐方法，其他修改方式在严格模式下将会报错。该方法只能进行同步操作，且方法名只能全局唯一。操作之中会有一些 hook 暴露出来，以进行 state 的监控等。

（6）State：页面状态管理容器对象。集中存储 Vue Components 中 data 对象的零散数据，全局唯一，以进行统一的状态管理。页面显示所需的数据从该对象中进行读取，利用 Vue 的细粒度数据响应机制进行高效的状态更新。

（7）Getters：state 对象读取方法。图中没有单独列出该模块，应该被包含在了 render 中，Vue Components 通过该方法读取全局 state 对象。

使用 Vuex 会有一定的门槛和复杂性，它的主要使用场景是大型单页应用，更适合多人协同开发，如果项目不是很复杂，或者希望短期内见效，需要认真考虑是否真的有必要使用 Vuex，也许 bus 方法就能很简单地解决你的需求。当然，并不是所有大型多人协同开发的 SPA 项目都必须使用 Vuex，事实上，我们在一些生产环境中只使用 bus 也能实现得很好，用与否主要取决于团队和技术储备。

8.4.2 基本用法

1. 安装

首先通过 NPM 安装 Vuex 依赖,命令如下:

```
npm install vuex -- save
```

2. 使用

(1) 在项目根目录创建 store 文件夹,在该文件夹下创建 index.js,代码如下:

```
//store/index.js
import Vue from 'vue'
import Vuex from 'vuex'
//挂载 Vuex
Vue.use(Vuex)
//创建 VueX 对象
const store = new Vuex.Store({
    state:{
        //存放的键值对就是所要管理的状态
        name:'这是 Vuex 的第一个数据'
}})
//导出 store
export default store
```

(2) 在 main.js 文件中将 store 挂载到当前项目的 Vue 实例中,代码如下:

```
//main.js 部分代码省略
import store from './store'//导入 store,自动会寻找 index.js
new Vue({
  el:'#app',
  router,
  store, //store:store 和 router 一样,将我们创建的 Vuex 实例挂载到这个 Vue 实例中
components: { App },
  template: '<App/>'
})
```

在组件中使用 Vuex,如在 HelloWorld.vue 中,我们要将 state 中定义的 name 在 h1 标签中显示,代码如下:

```
<template>
    <div id="hello">
        name:
    <h1>{{ $store.state.name }}</h1><!-- 获取 state 中的值 -->
```

```
        </div>
    </template>
```

或者在组件方法中使用：

```
//...
methods:{
    add(){
        console.log(this.$store.state.name)
}},
```

3. Vuex中的核心内容

在Vuex对象中，其实不止有state，还有用来操作state中数据的方法集。成员列表如下。

（1）State：数据源存放状态。

（2）mutations state：成员操作。

（3）Getters：加工state成员给外界。

（4）Actions：异步操作。

（5）Modules：模块化状态管理。

Mutations：mutations是操作state数据的方法的集合，例如对该数据的修改、增加、删除等。

mutations方法有默认的形参([state],[payload])：

（1）state是当前Vuex对象中的state。

（2）payload是该方法在被调用时传递参数使用的。

例如编写一个方法，当被执行时，把state中管理的name值修改成"beixi"，代码如下：

```
//store/index.js 部分代码省略
const store = new Vuex.Store({
  state:{
        name:'helloVueX'    //存放的键值对就是所要管理的状态
  },
  mutations:{
    edit(state){            //ES6语法,等同edit:funcion(){...}
      state.name = 'beixi'
    }
  }
})
```

而在组件中，我们需要调用mutation，例如在HelloWorld.vue的任意方法中调用，代码如下：

```
//HelloWorld.vue 部分代码省略
<button @click = "edit">获取修改后数据</button>
export default {
  methods: {
    edit(){
      this.$store.commit('edit')
    }
  }}
```

Mutation 传值，即在实际生产过程中，需要在提交某个 mutation 时携带一些参数供方法使用。

提交单个值时，代码如下：

```
this.$store.commit('edit',18)
```

当需要提交多参时，推荐把它们放在一个对象中提交，代码如下：

```
this.$store.commit('edit',{age:18,sex:'男'})
```

在 HelloWorld.vue 页面方法中接收挂载的参数，代码如下：

```
edit(state,payload){
    state.name = 'beixi'
    console.log(payload)      //18 或{age:18,sex:'男'}
}
```

增删 state 中的成员。

为了配合 Vue 的响应式数据，我们在 Mutations 的方法中应当使用 Vue 提供的方法进行操作。如果使用 delete 或者以 xx.xx = xx 的形式去删或增，则 Vue 不能对数据进行实时响应。

Vue.set 为某个对象成员设置值，若不存在则新增。例如对 state 对象添加一个 age 成员，代码如下：

```
Vue.set(state,"age",18)
```

Vue.delete 删除成员，将刚刚添加的 age 成员删除，代码如下：

```
Vue.delete(state,'age')
```

Getters：相当于 Vue 中的 computed 计算属性，getter 的返回值会根据它的依赖被缓存起来，且只有当它的依赖值发生了改变时才会被重新计算，这里我们可以通过定义 Vuex

的 Getters 来获取，Getters 可以用于监听 state 中值的变化，返回计算后的结果。

Getters 中的方法有两个默认参数。

（1）state：当前 Vuex 对象中的状态对象。

（2）getters：当前 getters 对象，用于调用 getters 下的其他 getter 并使用，代码如下：

```
//store/index.js 部分代码省略
const store = new Vuex.Store({
//...
  getters:{
    nameInfo(state){
      return "姓名:" + state.name
    },
    fullInfo(state,getters){
      return getters.nameInfo + '年龄:' + state.age
    }
  }
})
```

HelloWorld.vue 组件中调用，代码如下：

```
<div id="hello">
    <h3>从 Getters 获取计算后的值：{{this.$store.getters.fullInfo}}</h3>
</div>
```

Actions：由于直接在 mutation 方法中进行异步操作，会引起数据失效。所以提供了 Actions，用来专门进行异步操作，最终提交 mutation 方法。

Actions 中的方法有两个默认参数。

（1）context：上下文（相当于箭头函数中的 this）对象。

（2）payload：挂载参数。

例如我们在两秒后执行 mutations 中的 edit 方法，由于 setTimeout 是异步操作，所以需要使用 actions，代码如下：

```
//store/index.js 部分代码省略
const store = new Vuex.Store({
//...
  actions:{
    aEdit(context,payload){
      setTimeout(()=>{
        context.commit('edit',payload)
      },2000)
    }
  }
})
```

在 HelloWorld.vue 组件中调用,代码如下:

```
//HelloWorld.vue 部分代码省略
<button @click = "aEdit">异步获取数据</button>
export default {
  methods: {
    aEdit(){
      this.$store.dispatch('aEdit',{age:18})
    }
  }}
```

对上述代码进行改进,由于是异步操作,所以我们可以将异步操作封装为一个 Promise 对象,代码如下:

```
aEdit(context,payload){
    return new Promise((resolve,reject) =>{
        setTimeout(() =>{
            context.commit('edit',payload)
            resolve()
        },2000)
    })
}
```

8.4.3 模块组

当项目庞大且状态非常多时,可以采用模块化管理模式。Vuex 允许我们将 store 分割成模块(module)。每个模块拥有自己的 state、mutation、action、getter 甚至嵌套子模块——从上至下进行同样方式的分割。

1. 大致的结构

模块组大致结构如下:

```
//模块
Aconst moduleA = {
  state: { ... },
  mutations: { ... },
  actions: { ... },
  getters: { ... }}
//模块
Bconst moduleB = {
  state: { ... },
  mutations: { ... },
  actions: { ... }}
//组装
```

```
const store = new Vuex.Store({
  modules: {
    a: moduleA,
    b: moduleB
  }})
//取值
store.state.a // -> moduleA 的状态
store.state.b // -> moduleB 的状态
```

2. 详细示例

(1) 首先在 store/index.js 文件夹中新建结果模块,代码如下:

```
import Vue from 'vue'
import Vuex from 'vuex'
Vue.use(Vuex)
const moduleA = {
  state: {
    ceshi: "这里是moduleA"
  },
  mutations: {}
};
const moduleB = {
  state: {
    ceshi2: "这里是moduleB",
  }
};
export default new Vuex.Store({
  modules: {
    a: moduleA,          //访问该模块 this.$store.state.a
    b: moduleB
  }
})
```

(2) 在 router/index.js 中配置路由,代码如下:

```
//router/index.js    部分代码省略
import data from '@/components/data'
export default new Router({
    routes: [{
    path: '/data',
    name: 'data',
    component: data
  }]
})
```

(3) 然后在 data.vue 组件中进行调用，代码如下：

```
<!-- 如果我们引入的模块是 A,并且想要获取 moduleA 中的 ceshi 的值,我们可以这样实现 -->
<template>
    <div>
        {{ceshi}}
    </div>
</template>
<script>
  export default {
    name: 'App',
    computed: {
      ceshi() {
        return this.$store.state.a.ceshi
      }
    }
  }
</script>
```

第 9 章 实战：百度音乐项目

本章将结合书中提到的知识点开发一个具有代表性的移动端百度音乐项目。百度音乐是利用目前流行的 Vue 框架和 Axios 进行跨域请求百度音乐接口、结合 swiper 来布局的比较综合性的前端项目，可以让读者提前体验前后端分离的魅力。

9.1 音乐列表

1. 创建项目

首先利用脚手架构建 Vue 项目（如：vueMusic），通过 NPM 安装 Axios 插件。并在 main.js 中导入并初始化配置：

```
//main.js
import Vue from 'vue'
import App from './App'
import router from './router'
import axios from 'axios'
//加载到原型上
Vue.prototype.$axios = axios;
Vue.config.productionTip = false
new Vue({
  el: '#app',
  router,
  components: { App },
  template: '<App/>'
})
```

2. 调整项目结构

调整项目结构，删除 HelloWorld.vue 组件及相关配置。components 用于存放公共的组件页面，在 src 下新建 pages 文件夹用于存放页面组件。在 App.vue 中设置初始化样式，代码如下：

```
<style>
 *{margin:0;padding:0;-webkit-tap-highlight-color:rgba(0,0,0,0);-webkit-text-
size-adjust:none;-webkit-user-select:none;-ms-user-select:none;user-select:none;
font-family:Arial,"微软雅黑";}
   img{border:none;max-width:100%;vertical-align:middle;} body,p,form,input,button,dl,
dt,dd,ul,ol,li,h1,h2,h3,h4,h5,h6{margin:0;padding:0;list-style:none;overflow-x:hidden}
   h1,h2,h3,h4,h5,h6{font-size:100%;} input,textarea{-webkit-user-select:text;-ms-
user-select:text;user-select:text;-webkit-appearance:none;font-size:1em;line-
height:1.5em;}
   table{border-collapse:collapse;}
   input,select,textarea{outline:none;border:none;background:none;}
   a{outline:0;cursor:pointer;*star:expression(this.onbanner=this.blur());}
   a:link,a:active{text-decoration:none;}
   a:visited{text-decoration:none;}
   a:hover{color:#f00;text-decoration:none;}
   a{text-decoration:none;-webkit-touch-callout:none;}
   em,i{font-style:normal;}
   li,ol{list-style:none;}
   html{font-size:10px;}
   .clear{clear:both;height:0;font-size:0;line-height:0;visibility:hidden;overflow:
hidden;}
   .fl{float:left;}
   .fr{float:right;}
   body{margin:0 auto;max-width:640px;min-width:320px;color:#555;background:#f9f9f9;
height:100%;}
   a{color:#222;text-decoration:none}
   .active{      /*路由高亮*/
     color:red;
   }
</style>
```

3. 创建 MusicList 页面

在 components 目录下建立公共组件 MusicList.vue，用于存放歌曲列表，代码如下：

```
//components/MusicList.vue
<template lang="html">
  <div class="musiclist">
    <div class="panel hotsongs on">
      <ul class="list">
<!--静态歌曲列表-->
        <li class="song">
          <div class="poster">
            <img src="http://qukufile2.qianqian.com/data2/pic/b733a1a9fc0f63c7015be-
29b7b840b66/672866107/672866107.jpg@s_2,w_150,h_150" alt="桥边姑娘">
```

```html
        </div>
          <div class = "info">
            <div class = "name">桥边姑娘</div>
            <div class = "author">舞蹈女神诺涵</div>
          </div>
        </li>
      </ul>
    </div>
  </div>
</template>
<script>
  export default {
    name:"musiclist"
  }
</script>
<style scoped>
<!-- 由于样式较多,大家可自行下载源码查看,源码地址在书的开始 -->
</style>
```

4. 创建首页组件

在 pages 目录下新建首页组件 index.vue, 初始化并配置路由, 代码如下:

```
//src/router/index.js
import Vue from 'vue'
import Router from 'vue-router'
import index from '@/pages/index'
Vue.use(Router)
export default new Router({
  routes: [
    {
      path: '/',
      name: 'index',
      component: index
    }
  ]
})
```

将 MusicList.vue 组件导入首页中, 代码如下:

```
//src/pages/index.vue
<template>
<div>
<Musiclist></Musiclist>
</div>
```

```
</template>
<script>
    /*导入歌曲列表组件*/
    import Musiclist from "../components/MusicList";
    export default {
        name: "index",
        components:{
            Musiclist
        }
    }
</script>
```

页面显示效果如图 9-1 所示。

图 9-1　静态歌曲列表

9.1.1　跨域配置

如果希望实时动态地获取百度音乐中的歌曲列表,则需要使用 Axios 跨域请求百度音乐接口中的歌曲列表。百度音乐接口 API：https://www.jianshu.com/p/e9d43d15f6ba。跨域请求的配置代码如下：

```
//config/index.js 部分代码省略
proxyTable: {
    "/api": {
        target: "http://tingapi.ting.baidu.com",
        changeOrigin: true,
        pathRewrite: {
            '^SymbolYCp/api': ''
        }
    }
```

接着把网络基本 URL 挂载到 main.js 中,并且配置拦截器,代码如下：

```
//main.js 省略部分代码
/*qs 是一个增加了安全性的查询字符串解析和序列化字符串的库。*/
```

```
import qs from "qs"
/* 基本网络 URL 的挂载 */
Vue.prototype.HOST = '/api'
// 添加请求拦截器
axios.interceptors.request.use(function (config) {
  if(config.method === "post"){
    config.data = qs.stringify(config.data)
  }
  return config;
}, function (error) {
  // 对请求错误做些什么
  return Promise.reject(error);
});
// 添加响应拦截器
axios.interceptors.response.use(function (response) {
  // 对响应数据做点什么
  return response;
}, function (error) {
  // 对响应错误做点什么
  return Promise.reject(error);
});
```

接着修改 MusicList.vue 组件，使其能动态获取百度音乐接口中歌曲列表数据，代码如下：

```
//components/MusicList.vue 省略样式代码
<template lang = "html">
  <div class = "musiclist">
    <div class = "panel hotsongs on">
      <ul class = "list">
        <li class = "song" v-for = "(item,index) in musicData.song_list">
          <div class = "poster">
            <img :src = "item.pic_big" :alt = "item.title">
          </div>
          <div class = "info">
            <div class = "name">{{ item.title }}</div>
            <div class = "author">{{ item.author }}</div>
          </div>
        </li>
      </ul>
    </div>
  </div>
</template>
<script>
```

```
    export default {
      name:"musiclist",
      data(){
        return{
          loading:true,
          musicData:{
            song_list:[]
          }
        }
      },
      created(){
/* 参数：
        type = 1-新歌榜,2-热歌榜,11-摇滚榜,12-爵士,16-流行,21-欧美金曲榜,22-经
典老歌榜,23-情歌对唱榜
        size = 10 //返回条目数量
        offset = 0 //获取偏移
        */
        const musiclistUrl = this.HOST + "/v1/restserver/ting?method=baidu.ting.
                billboard.billList&type=1&size=5&offset=0"
        this.$axios.get(musiclistUrl)
          .then(res => {
            this.musicData = res.data
            this.loading = false
          })
          .catch(error =>{
            console.log(error);
          })
      }
    }
</script>
```

musiclistUrl 跨域请求百度音乐数据,type 代表音乐类型,size 代表获取歌曲数目。上述代码中 type 的值是固定不变的,我们也可以动态地获取对应类型的歌曲列表,代码如下：

```
//pages/index.vue 部分代码省略
<div>
    <!-- musictype 传入音乐类型参数 -->
    <Musiclist musictype="1"></Musiclist>
</div>

//components/MusicList.vue 部分代码省略
props:{//接收参数
        musictype:{
            type:String,
```

```
      default:"1"
    }
  },
  created(){
    /*注意将参数 type 值修改为动态值*/
    const musiclistUrl = this.HOST + "/v1/restserver/ting?method=baidu.ting.billboard.
            billList&type=" + this.musictype + "&size=5&offset=0"
    this.$axios.get(musiclistUrl)
      .then(res => {
        this.musicData = res.data
        this.loading = false
      })
      .catch(error =>{
        console.log(error);
      })
  }
```

9.1.2 音乐列表导航栏

9.1.1 节中我们通过手动传参更换音乐类型从而获取对应的歌曲列表,为了提高用户体验需要设置导航栏进行操作。在 pages 下新建 MusicNav.vue 导航栏组件,代码如下:

```
//pages/MusicNav.vue
<template lang="html">
  <div class="bars">
    <div class="bar on">
      <router-link to="/hot">热歌榜</router-link>
      <i></i><span class="gap-line"></span></div>
    <div class="bar">
      <router-link to="/new">新歌榜</router-link>
      <i></i><span class="gap-line"></span></div>
    <div class="bar">
      <router-link to="/king">King 榜</router-link>
      <i></i></div>
  </div>
</template>
<script>
export default {
}
</script>
<style scoped>
<!-- 由于样式较多,大家可自行下载源码查看,源码地址在书的开始 -->
</style>
```

在首页 index.vue 组件中把歌曲列表 MusicList.vue 组件替换为新建的导航栏组件,代码如下：

```
//pages/index.vue
<template>
    <div>
        <MusicNav></MusicNav>
        <!-- 挂载3个子组件 -->
        <router-view></router-view>
    </div>
</template>
<script>
  import MusicNav from "./MusicNav";
    export default {
        name: "index",
        components:{
            MusicNav
        }
    }
</script>
```

创建导航栏下的3个子页面，即热歌榜（HotMusic.vue）、新歌榜（NewMusic.vue）、King榜（KingMusic.vue），代码如下：

```
//pages/musicList/HotMusic.vue
<template lang="html">
    <div>
        <MusicList musictype="1" />
    </div>
</template>
<script>
import MusicList from "../../components/MusicList"
export default {
  name:"hot",
  components:{
    MusicList
  }
}
</script>

//pages/musicList/NewMusic.vue
<template lang="html">
    <div>
        <MusicList musictype="22" />
    </div>
</template>
```

```
<script>
import MusicList from "../../components/MusicList"
export default {
  name:"new",
  components:{
    MusicList
  }
}
</script>

//pages/musicList/KingMusic.vue
<template lang="html">
    <div>
        <MusicList musictype="2" />
    </div>
</template>
<script>
import MusicList from "../../components/MusicList"
export default {
  name:"king",
  components:{
    MusicList
  }
}
</script>
```

最后进行路由配置,代码如下:

```
//router/index.js 部分代码省略
import HotMusic from "../pages/musicList/HotMusic"
import NewMusic from "../pages/musicList/NewMusic"
import KingMusic from "../pages/musicList/KingMusic"
export default new Router({
  routes: [
    {
      path: '/',
      name: 'index',
      component: index,
      redirect:"/hot",
      children:[
        {
          path: '/hot',
          name: 'HotMusic',
          component: HotMusic
```

```
      },
      {
        path: '/king',
        name: 'KingMusic',
        component: KingMusic
      },
      {
        path: '/new',
        name: 'NewMusic',
        component: NewMusic
      }
    ]
  }
 ]
})
```

页面显示效果如图 9-2 所示。

图 9-2　加入导航页面效果

9.2　歌手信息

歌手信息一栏用于展示当前比较火热的歌手信息列表，首先我们创建静态热门歌手信息模型组件 SingerList.vue，主要用于存放公共歌手列表信息，代码如下：

```
//components/SingerList.vue
<template lang="html">
    <div class="container">
```

```html
            <div class="wrapper">
              <!-- 静态歌手信息列表 -->
              <div class="card">
                <div class="card-slider">
                  <div class="poster">
                    <img src="http://qukufile2.qianqian.com/data2/pic/140d09665d1c204efe00-973c3e16282c/612757841/612757841.jpg@s_2,w_240,h_240" alt="薛之谦">
                  </div>
                </div>
                <div class="info">
                  <div class="name">薛之谦</div>
                </div>
              </div>
            </div>
</template>
<script>
export default {
  name:"singerlist"
}
</script>
<style scoped>
<!-- 由于样式较多,大家可自行下载源码查看,源码地址在书的开始 -->
</style>
```

接着将歌手信息列表导入其对应的歌手组件Singer.vue中进行整个版块的布局,代码如下：

```html
//pages/Singer.vue
<template lang="html">
    <div class="mod-songlists">
        <div class="hd">
            <h2>热门歌手</h2>
        </div>
        <SingerList></SingerList>
    </div>
</template>
<script>
    /*导入歌手列表*/
import SingerList from "../components/SingerList"
export default {
        name:"singer",
        components:{
            SingerList
```

```
    }
  }
</script>
<style scoped>
<!-- 由于样式较多,大家可自行下载源码查看,源码地址在书的开始 -->
</style>
```

将热门歌手版块挂载到首页显示,代码如下:

```
//pages/index.vue 部分代码省略
<div>
    <Singer></Singer>
</div>
import Singer from "./Singer";
  export default {
      name: "index",
      components:{
          Singer
      }
  }
```

在 SingerList.vue 组件中复制多个静态歌手信息列表,页面显示效果如图 9-3 所示。

静态模型成型后,接着动态设置歌手信息列表,在百度音乐接口 API 中获取歌手信息,如图 9-4 所示,我们只需在百度音乐中手动获取热门歌手的 tinguid。

图 9-3 静态热门歌手　　　　图 9-4 百度音乐接口歌手信息

将获取的热门歌手的 tinguid 放在 Singer.vue 组件数组中遍历,并传参到歌手组件 SingerList.vue 中,动态加载歌手信息,代码如下:

```
//pages/Singer.vue    部分样式代码省略
<template lang = "html">
    <div class = "mod - songlists">
```

```html
            <div class = "hd">
                <h2>热门歌单</h2>
            </div>
            <!-- 统一的父级容器从 SingerList.vue 转移到这里 -->
            <div class = "container">
                <div class = "wrapper">
                    <!-- singerid 是歌手的对应 tinguid 值,向 SingerList.vue 组件传参 -->
                    <SingerList :key = "index" v - for = "(item,index) in singerListData" :singerid = "item" />
                </div>
            </div>
        </div>
    </div>
</template>
<script>
    import SingerList from "../components/SingerList"
    export default {
        name:"singer",
        data(){
            return{
                /* 手动在百度音乐中获取热门歌手的 tinguid 值 */
                singerListData:["2517","1097","1376","1557","1052","1133"]
            }
        },
        components:{
            SingerList
        }
    }
</script>
<style scoped>
    .wrapper{
        border: none;
        margin: 0 - 5px;
        overflow: hidden;
        position: relative;
        width: auto;
    }
</style>

//components/SingerList.vue    样式代码省略
<template lang = "html">
    <div class = "card">
        <div class = "card - slider">
            <div class = "poster">
                <img :src = "singerInfo.avatar_big" :alt = "singerInfo.name">
            </div>
        </div>
        <div class = "info">
```

```
        <div class = "name">{{ singerInfo.name }}</div>
      </div>
    </div>
</template>
<script>
  export default {
    name:"singerlist",
    data(){
      return{
        singerInfo:{}
      }
    },
    props:{//接收 Singer.vue 组件传过来的歌手 singerid 参数
      singerid:{
        type:String,
        default:"0"
      }
    },
    created(){
      const SingerUrl = this.HOST + "/v1/restserver/ting?method = baidu.ting.artist.
              getInfo&tinguid = " + this.singerid
      this.$axios.get(SingerUrl)
        .then(res => {
          this.singerInfo = res.data
        })
        .catch(error => {
          console.log(error);
        })
    }
  }
</script>
```

页面显示效果如图 9-5 所示。

图 9-5　动态展示热门歌手

当单击歌手头像及其信息时，需要显示该歌手所对应的歌曲列表。给歌手信息设置路由导航，代码如下：

```html
//components/SingerList.vue    部分代码省略
<template lang="html">
    <!-- 动态接收路由参数 -->
    <router-link :to="{name:'SingerInfo',params:{singerid:singerid}}" tag="div" class="card">
        <div class="card-slider">
            <div class="poster">
                <img :src="singerInfo.avatar_big" :alt="singerInfo.name">
            </div>
        </div>
        <div class="info">
            <div class="name">{{ singerInfo.name }}</div>
        </div>
    </router-link>
</template>
```

接着创建承载歌曲列表页面 SingerInfo.vue，并进行路由配置，代码如下：

```html
//pages/Singer/SingerInfo.vue
<template lang="html">
    <div class="singerinfo">
        <SingerMusicList />
    </div>
</template>
<script>
/* 导入歌手歌曲 */
import SingerMusicList from "../../components/SingerMusicList"
export default {
  name:"singerinfo",
  components:{
    SingerMusicList
  }
}
</script>

//router/index.js 部分代码省略
import SingerInfo from "../pages/Singer/SingerInfo"
export default new Router({
    routes:[
        {
            path:"/singer/:singerid",//在路由中传入单击歌手对应的 id
            name:"SingerInfo",
            component:SingerInfo
```

```
    }
  ]
})
```

最后将歌手信息页面导入的 SingerMusicList.vue 页面创建出来,用于展示歌手所对应的歌曲列表,代码如下:

```
//components/SingerMusicList.vue
<template lang="html">
  <div class="musiclist">
    <div class="panel hotsongs on">
      <ul class="list">
        <li class="song" v-for="(item,index) in musicData.songlist">
          <div class="poster">
            <img :src="item.pic_big" :alt="item.title">
          </div>
          <div class="info">
            <div class="name">{{ item.title }}</div>
            <div class="author">{{ item.author }}</div>
          </div>
        </li>
      </ul>
    </div>
  </div>
</template>
<script>
export default {
  name:"singermusiclist",
  data(){
    return{
      musicData:{
        songlist:[]
      }
    }
  },
  created(){
                  /*注意这里获取的是歌手歌曲列表接口*/
    const musiclistUrl = this.HOST + "/v1/restserver/ting?method=baidu.ting.artist.getSongList&tinguid=" + this.$route.params.singerid + "&limits=20&use_cluster=1&order=2"
    this.$axios.get(musiclistUrl)
    .then(res => {
      this.musicData = res.data
    })
    .catch(error =>{
```

```
            console.log(error);
        })
    }
}
</script>
<style scoped>
<!-- 由于样式较多,大家可自行下载源码查看,源码地址在书的开始 -->
</style>
```

这样在单击歌手时,就会展示其对应的歌曲列表信息。

9.3 歌曲播放

当单击歌名(包括歌手下的音乐列表)时,应该跳转并携带歌曲的 songid 值到歌曲播放页面,以便播放对应的歌曲。创建歌曲播放页面 MusicPlay.vue,并且进行路由配置,代码如下:

```
//pages/MusicPlay.vue
<template lang="html">
    <div>
    歌曲播放
        {{$route.params.songid}} <!-- 获取路由中传来的歌曲 id -->
    </div>
</template>
<script>
export default {
  name:"play",
  }
</script>

//router/index.js   部分代码省略
{
  path:"/play/:songid",//从音乐列表页面携带来的歌曲 id
  name:"MusicPlay",
  component:MusicPlay
}
```

我们单击的音乐列表所在的页面是 MusicList.vue 和 SingerMusicList.vue,要想单击音乐跳转到播放页面就需要把<li>替换成路由跳转标签<router-link>,并且携带歌曲 songid 值到歌曲播放页面,代码如下:

```
//components/MusicList.vue   部分代码省略
<ul class="list">
```

```
    <!-- 单击歌曲列表跳转并携带歌曲 id 到 MusicPlay 路由 -->
    <router-link :key="index" :to="{name:'MusicPlay',params:{songid:item.song_id}}" tag=
"li" class="song" v-for="(item,index) in musicData.song_list">
        <div class="poster">
            <img :src="item.pic_big" :alt="item.title">
        </div>
        <div class="info">
            <div class="name">{{ item.title }}</div>
            <div class="author">{{ item.author }}</div>
        </div>
    </router-link>
</ul>

//components/SingerMusicList.vue   部分代码省略
<ul class="list">
    <!-- 单击歌曲列表跳转并携带歌曲 id 到 MusicPlay 路由 -->
    <router-link :key="index" :to="{name:'MusicPlay',params:{songid:item.song_id}}" tag=
"li" class="song" v-for="(item,index) in musicData.songlist">
        <div class="poster">
            <img :src="item.pic_big" :alt="item.title">
        </div>
        <div class="info">
            <div class="name">{{ item.title }}</div>
            <div class="author">{{ item.author }}</div>
        </div>
    </router-link>
</ul>
```

这样我们在单击任意一首歌时都能跳转到歌曲播放页面,并能成功地接收歌曲 id 值。在 MusicPlay.vue 页面,根据得到的动态歌曲 id 值可访问百度音乐歌曲接口,从而获取歌曲进而播放,代码如下:

```
//pages/MusicPlay.vue
<template lang="html">
    <div>
    <!-- file_link 是获取的歌曲数据 -->
        <audio ref="player" :src="playData.bitrate.file_link" controls>
        </audio>
    </div>
</template>
<script>
export default {
  name:"play",
  data(){
```

```
        return{
          playData: {
            bitrate:{
              file_link:"" //需初始化这个值,否则报错并提示file_link没有定义
            }
          }
        }
      },
      created(){
        /*百度音乐歌曲播放接口,动态获取路由中的歌曲songid*/
        const playUrl = this.HOST + "/v1/restserver/ting?method=baidu.ting.song.play&songid=" + this.$route.params.songid
          this.$axios.get(playUrl)
            .then(res => {
              this.playData = res.data
            })
            .catch(error => {
              console.log(error);
            })
      }
    }
</script>
```

注意：关于使用file_link不能播放的问题,是因为百度使用Http中的Referer头字段来防止盗链,在index.html文件中加上<meta name="referrer" content="never">这一句让发送出去的Http包都不含Referer字段就可以了。

数据加载完毕后,渲染歌曲播放页面,并加入一些阿里巴巴矢量图标,代码如下：

```
//pages/MusicPlay.vue
<template lang="html">
  <div class="play">
    <div class="header">
      <div class="title">
        <router-link to="/"><!--单击回到首页-->
          <i class="iconfont icon-shouye left"></i>
        </router-link>
        <div class="music-info">
          <p>{{ playData.songinfo.title }}</p>
          <p>{{ playData.songinfo.author }}</p>
        </div>
        <router-link to="/"><i class="iconfont icon-sousuo
                    right"></i></router-link>
      </div>
    </div>
```

```html
        <div class = "song-info">
          <div class = "song-info-img">
            <img :src = "playData.songinfo.pic_big">
          </div>
          <div class = "iconbox">
            <i class = "iconfont icon-shoucang2 left"></i>
            <i class = "box"></i>
            <i class = "iconfont icon-xiazai right"></i>
          </div>
        </div>
        <div class = "song">
          <audio :src = "playData.bitrate.file_link" controls>
          </audio>
        </div>
      </div>
    </div>
</template>
<script>
  /*导入阿里巴巴矢量图标*/
  import "../assets/font/iconfont.css"
export default {
  name:"play",
  data(){
    return{
      playData: {
        songinfo:{//需定义,否则报 songinfo.title 没有定义

},
        bitrate:{
          file_link:"" //需初始化这个值,否则报错并提示 file_link 没有定义
}
      }
    }
  },
  created(){
    /*百度音乐歌曲播放接口,动态获取路由中的歌曲 songid*/
    const playUrl = this.HOST + "/v1/restserver/ting?method = baidu.ting.song.play&songid = " + this.$route.params.songid
    this.$axios.get(playUrl)
      .then(res => {
        this.playData = res.data
      })
      .catch(error => {
        console.log(error);
      })
  }
```

```
        }
    </script>
    <style scoped>
    <!-- 由于样式较多,大家可自行下载源码查看,源码地址在书的开始 -->
    </style>
```

页面显示效果如图 9-6 所示。

图 9-6 歌曲播放页面展示

9.4 轮播图

在首页顶部使用 swiper 实现轮播图的无缝滚动。首先在项目下安装 swiper,命令如下:

```
npm install vue-awesome-swiper --save
```

在 main.js 中引入 swiper 及其样式,代码如下:

```
import VueAwesomeSwiper from 'vue-awesome-swiper'
import 'swiper/dist/css/swiper.css'
Vue.use(VueAwesomeSwiper)
```

接着创建公共轮播图 SwiperSilder.vue 组建页面,代码如下:

```
//components/SwiperSilder.vue
<template lang="html">
  <div class="silder">
    <swiper :options="swiperOption">
      <swiper-slide>
```

```html
            <img src="../assets/banner/b1.jpg" alt="">
        </swiper-slide>
        <swiper-slide>
            <img src="../assets/banner/b2.jpg" alt="">
        </swiper-slide>
        <swiper-slide>
            <img src="../assets/banner/b3.jpg" alt="">
        </swiper-slide>
        <div class="swiper-pagination" slot="pagination"></div>
    </swiper>
  </div>
</template>
<script>
export default {
  name:"slider",
  data(){
    return{
      swiperOption: {
        pagination: {//设置分页器
el: '.swiper-pagination',
        },
        loop:true
      }
    }
  }
}
</script>
```

接着将 SwiperSilder.vue 导入首页, 使其在该页面展示, 代码如下:

```html
//pages/index.vue 部分代码省略
<template>
    <div>
        <SwiperSlider />
    <!-- .... -->
    </div>
</template>
<script>
    import SwiperSlider from "../components/SwiperSilder"
    export default {
        name: "index",
      components:{
        SwiperSlider
      }
    }
</script>
```

页面显示效果如图 9-7 所示。

轮播图效果完成后,在热门榜单下方添加一个查看更多榜单的功能,供大家查看更多的歌单。首先在 MusicList.vue 页面添加查看更多榜单功能,并携带 musictype 参数给榜单详情 Recommend 路由,使其跳转到相对应的榜单详情页面,代码如下:

图 9-7　页面顶部轮播图展示

```
//components/MusicList.vue    部分代码省略
< div class = "panel hotsongs on">
    <!-- ... -->
    <!-- 跳转到详情列表页,并携带 musictype 参数给 Recommend 路由,使其跳转到相对应的榜单页面 -->
    < router - link :to = "{name:'Recommend',params:{musictype:musictype}}" tag = "div" class = "more - songs">
        查看更多榜单 &gt;
    </ router - link >
</div >
< style >
.more - songs{
    color: #999;
    margin - top: 9px;
    font - size: 12px;
    text - align: center;
    height: 32px;
    line - height: 32px;
}
</style >
```

页面显示效果如图 9-8 所示。

图 9-8　查看更多绑定功能

创建榜单详情页面,在接收到路由传来的音乐类型参数后,调用歌曲列表接口查询更多歌单,并且为歌单设置路由跳转到播放页面,代码如下：

```html
//pages/Recommend/Recommend.vue
<template lang="html">
  <div class="recommend">
    <div class="panel hotsongs">
      <ul class="list">
        <!-- 单击每首歌跳转到播放页面,并携带 songid 给路由 --><router-link :key="index" :to="{name:'MusicPlay',params:{songid:item.song_id}}" tag="li" class="song" v-for="(item,index) in musicData">
          <div class="poster">
            <img :src="item.pic_big" :alt="item.title">
          </div>
          <div class="info">
            <div class="name">{{ item.title }}</div>
            <div class="author">{{ item.author }}</div>
          </div>
        </router-link>
      </ul>
    </div>
  </div>
</template>
<script>
  export default {
    name:"recommendlist",
    data(){
      return{
        musicData:[],
      }
    },
    created(){
      /*调用歌曲列表接口*/
      const musiclistUrl = this.HOST + "/v1/restserver/ting?method=baidu.ting.billboard.billList&type=" + this.$route.params.musictype + "&size=20&offset=" + this.offsetNum
      this.$axios.get(musiclistUrl)
        .then(res => {
          this.musicData = this.musicData.concat(res.data.song_list)
        })
        .catch(error =>{
          console.log(error)
        })
    }
  }
```

```
</script>
<style scoped>
<!-- 由于样式较多,请读者自行下载源代码查看,源代码扫描前言中的二维码即可下载 -->
</style>

//router/index.js    部分代码省略
import Recommend from "../pages/Recommend/Recommend"
export default new Router({
  routes:[
    {
      path:"/recommend/:musictype",//接收 MusicList.vue 传来的 musictype 参数值
name:"Recommend",
      component:Recommend
    }
  ]
})
```

9.5 搜索实现

在轮播图顶部添加搜索功能,创建搜索页面 Search.vue,并将其添加到首页顶部,代码如下:

```
//components/Search.vue
<template lang="html">
  <div class="search">
    <div class="xw a2" style="position: relative;">
      <div class="y1 hasCancel">
        <i id="iconToSearch" class="y2 y4"></i>
        <input v-model="searchdata" type="search" style="max-width:90%" class="y0" placeholder="搜索你喜欢的音乐">
        <i id="iconToClear" class="a1q y4"></i>
      </div>
      <!-- 单击跳转到搜索详情页 SearchView.vue,并将搜索内容传递给 SearchView 路由 -->
      <router-link tag="span" :to=""{name:'SearchView',params:{searchcontent:searchdata}}" class="y5">搜索</router-link>
    </div>
  </div>
</template>
<script>
  export default {
    name:"search",
    data(){
      return{
```

```
            searchdata:""
        }
    }
}
</script>
<style scoped>
<!-- 由于样式较多,请读者自行下载源代码查看,源代码扫描前言中的二维码即可下载 -->
</style>

//pages/index.vue    部分代码省略
<template>
<div>
<Search/><!-- 添加到首页顶部 -->
<!-- ... -->
</div>
</template>
<script>
    import Search from "../components/Search"
    export default {
        name: "index",
        components:{
            Search
        }
    }
</script>
```

页面显示效果如图9-9所示。

图9-9　搜索功能

创建搜索详情页面,并配置路由,代码如下:

```
//pages/search/SearchView.vue
<template lang = "html">
  <div class = "searchlist">
    <div class = "panel hotsongs">
      <ul class = "list">
<router - link :key = "index" :to = "{name:'MusicPlay',params:{songid:item.songid}}" tag = "li" class = "song" v - for = "(item,index) in musicData.song">
```

```html
            <div class="info">
              <div class="name">{{ item.songname }}</div>
              <div class="author">{{ item.artistname }}</div>
            </div>
        </router-link>
      </ul>
    </div>
  </div>
</template>
<script>
  export default {
    name:"searchlist",
    data(){
      return{
        musicData:{
        }
      }
    },
    created(){
      /*调用搜索接口,接收路由传来的搜索内容*/
      const searchUrl = this.HOST + "/v1/restserver/ting?method=baidu.ting.search.catalogSug&query=" + this.$route.params.searchcontent
      this.$axios.get(searchUrl)
        .then(res => {
          this.musicData = res.data
        })
        .catch(error => {
          console.log(error);
        })
    }
  }
</script>
<style scoped>
<!-- 由于样式较多,请读者自行下载源代码查看,源代码扫描前言中的二维码即可下载 -->
</style>

//router/index.js      部分代码省略
import SearchView from "../pages/search/SearchView"
export default new Router({
  routes: [
    {
      path:"/searchview/:searchcontent",//接收 Search.vue 传来的搜索内容
      name:"SearchView",
      component:SearchView
    }
  ]
})
```

Spring Boot 篇

第 10 章　进入 Spring Boot 世界
第 11 章　Spring Boot 整合 Web 开发
第 12 章　应用开发
第 13 章　Spring Boot 热部署和 Postman 工具
第 14 章　Spring Boot 整合数据库
第 15 章　Spring Boot 整合持久层技术
第 16 章　Spring Boot 安全框架
第 17 章　项目构建与部署
第 18 章　部门管理系统

第 10 章 进入 Spring Boot 世界

Spring Boot 是一个框架，一种全新的编程规范，它的产生简化了框架的使用，所谓简化是指简化了 Spring 众多框架中所需的大量且烦琐的配置文件，所以 Spring Boot 是一个服务于框架的框架，服务范围是简化配置文件。从本质上来说，Spring Boot 其实就是 Spring 框架的另一种表现形式。

10.1 Spring Boot 简介

Spring 一直在飞速地发展，如今已经成为在 Java EE 开发中真正意义上的标准，但是随着技术的发展，Java EE 使用 Spring 逐渐变得笨重起来，大量的 XML 文件存在于项目之中。烦琐的配置，整合第三方框架的配置问题，导致了开发和部署效率降低。为了使开发者能够快速搭建 Java EE 项目，Spring Boot 应运而生。Spring Boot 是目前流行的微服务框架，使用 Spring Boot 可以快速创建基于 Spring 生产级的独立应用程序。Spring Boot 提供了很多核心功能，如自动化配置、starter 简化 Maven 配置、内嵌 Servlet 容器、应用监控等功能，它还集成了大量常用的第三方库配置，Spring Boot 应用中这些第三方库几乎可以开箱即用，使开发者能够更加专注于业务逻辑。概括来说，Spring Boot 主要有以下优势：

（1）快速创建独立运行的 Spring 项目及与主流框架集成。
（2）使用嵌入式的 Servlet 容器，应用无须打包成 War 包。
（3）starters 自动依赖与版本控制。
（4）大量的自动配置，从而简化开发，需要时也可修改默认值。
（5）无须配置 XML，无代码生成，开箱即用。
（6）在准生产环境运行时应用监控。
（7）与云计算的天然集成。

微服务是未来发展的趋势，项目会从传统架构慢慢转向微服务架构，因为微服务可以使不同的团队专注于更小范围的工作职责、使用独立的技术、更安全更频繁地部署。而 Spring Boot 继承了 Spring 的优良特性，与 Spring 一脉相承，而且支持各种 REST API 的实现方式。Spring Boot 也是官方大力推荐的技术，可以看出，Spring Boot 是未来发展的一个大趋势。

10.2　Spring Boot 环境准备

在开始学习 Spring Boot 前,应先准备好开发环境。本节将介绍如何安装 JDK、IntelliJ IDEA 及 Apache Maven。

10.2.1　JDK 环境

建议使用 JDK1.8 及以上的版本,官方下载地址为 https://www.oracle.com/java/technologies/javase/javase-jdk8-downloads.html。大家可以根据自己的需求选择合适的 JDK 安装包,这里不过多赘述。

10.2.2　开发工具 IDEA

IDEA,全称 IntelliJ IDEA,是 Java 语言的集成开发环境,IDEA 在业界被公认为最好的 Java 开发工具之一,尤其在智能代码助手、代码自动提示、重构、J2EE 支持、Ant、JUnit、CVS 整合、代码审查、创新的 GUI 设计等方面的功能可以说是超常的。

IntelliJ IDEA 在 2015 年的官网上这样介绍自己:IntelliJ IDEA 主要用于支持 Java、Scala、Groovy 等语言的开发工具,同时具备支持目前主流的技术和框架,擅长于企业应用、移动应用和 Web 应用的开发。

在 IntelliJ IDEA 的官方网站 https://www.jetbrains.com/idea/download/#section=Windows 下载 IDEA。大家可以根据自己的需求选择合适的 IDEA 版本,只要版本不过低即可。下载完成后,运行安装程序,根据提示安装即可。

10.2.3　安装与配置 Maven

Maven 是目前最流行的项目管理工具之一。Maven 官方下载网址为 http://maven.apache.org/download.cgi,大家可以根据自己的需求下载合适版本的 Maven,本书选用的版本为 apache-maven-3.3.9。虽然在 IDEA 中已经包含 Maven 插件,但是笔者还是希望大家在工作中能安装自己的 Maven 插件,以此方便项目配置。

下载完成后解压,然后将安装路径(例如 D:\java\apache-maven-3.3.9\bin)配置在环境变量 path 中。配置完成后,在命令窗口中执行:mvn -v,如果输出如图 10-1 所示的内容,表示 Maven 安装成功。

安装成功后,在 IDEA 中配置 Maven,具体步骤如下:

(1) 在安装目录(如:D:\java\apache-maven-3.3.9)下新建文件夹 repository,用来作为本地仓库。

(2) 修改 apache-maven-3.3.9→conf→settings 配置文件,如图 10-2 所示。

(3) 在 IDEA 界面中选择 File→Settings,在窗口中搜索 Maven 选项,配置为自己的 Maven,如图 10-3 所示。

第10章 进入Spring Boot世界 | 227

```
C:\WINDOWS\system32\cmd.exe

C:\Users\Administrator.SC-201803081145>mvn -v
Apache Maven 3.3.9 (bb52d8502b132ec0a5a3f4c09453c07478323dc5;
Maven home: D:\java\apache-maven-3.3.9
Java version: 1.8.0_131, vendor: Oracle Corporation
Java home: C:\Program Files\Java\jdk1.8.0_131\jre
Default locale: zh_CN, platform encoding: GBK
OS name: "windows 10", version: "10.0", arch: "amd64", family
```

图 10-1　Maven 安装成功

```
54        //第一处修改
55        -->
56    <localRepository>D:\java\apache-maven-3.3.9\repository</localRepository>

159       //第二处修改,添加阿里镜像
160       -->
161   <mirror>
162     <id>alimaven</id>
163     <name>aliyun maven</name>
164     <url>http://maven.aliyun.com/nexus/content/groups/public/</url>
165     <mirrorOf>central</mirrorOf>
166   </mirror>
```

图 10-2　settings.xml 配置文件修改

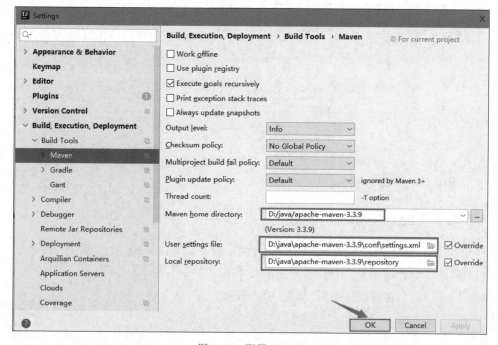

图 10-3　配置 Maven

10.3 Spring Boot 的三种创建方式

Spring Boot 本质上是一个 Maven 工程,下面介绍三种创建方式。

10.3.1 在线创建

在线创建是 Spring Boot 官方提供的一种创建方式,官网地址为 https://start.spring.io/,打开如图 10-4 所示页面。

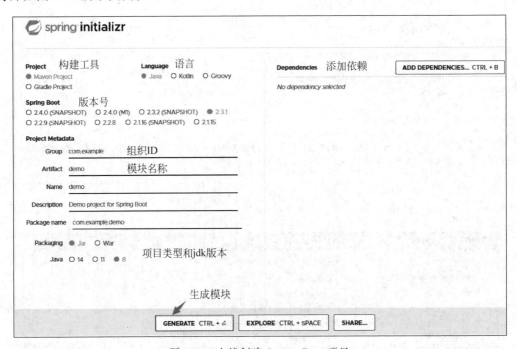

图 10-4 在线创建 Spring Boot 项目

在页面中将这些信息填写完成后,单击 GENERATE 按钮生成模块并下载到本地,解压后使用 IDEA 打开该项目即可使用。

10.3.2 通过 Maven 创建

使用 Maven 创建 Spring Boot 项目,以 IDEA 为开发工具,具体步骤如下:
(1) 创建一个普通 Maven 项目,如图 10-5 所示。
(2) 输入组织名称和模块名称,如图 10-6 所示。

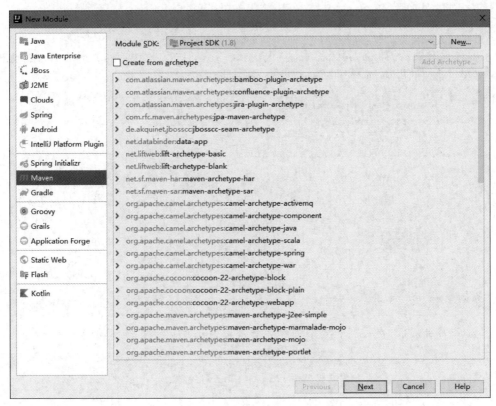

图 10-5　构建普通 Maven 项目

图 10-6　输入 Maven 信息

（3）创建完成后，在 pom.xml 文件中添加如下依赖。

```xml
<!-- Spring Boot 的父级依赖,只有继承它的项目才是 Spring Boot 项目。
    spring-boot-starter-parent 是一个特殊的 starter,它用来提供相关的 Maven 默认依赖。
使用它之后,常用的包依赖可以省去 version 标签。 -->
<parent>
    <groupId>org.springframework.boot</groupId>
    <artifactId>spring-boot-starter-parent</artifactId>
    <version>2.2.4.RELEASE</version>
    <relativePath/><!-- lookup parent from repository -->
</parent>
<dependencies>
<!-- Spring Boot 提供了非常多的 starter 启动器。spring-boot-starter-web 支持全栈式 Web
开发,包括 Tomcat 和 spring-webmvc。 -->
    <dependency>
        <groupId>org.springframework.boot</groupId>
        <artifactId>spring-boot-starter-web</artifactId>
    </dependency>
</dependencies>
```

（4）添加完成后，在 java 目录下创建一个名为 App 的启动类，代码如下：

```java
@EnableAutoConfiguration         //启用 Spring 的自动加载配置
@RestController                  //相当于 Controller + ResponseBody 注解
public class App {
    public static void main(String[] args) {
        SpringApplication.run(App.class, args);
    }
    @GetMapping("/hello")
    public String hello(){
        return "hello";
    }
}
```

最后执行 main 方法就可以启动一个 Spring Boot 项目了。

10.3.3 使用 Spring Initializer 快速创建

上述创建 Spring Boot 项目的两种方式都比较烦琐，推荐大家直接使用 Spring Initializer 的方式快速构建 Spring Boot 项目，本书后面的项目都将采用这种方式创建。这种方式不但可以生成完整的目录结构，还可以生成一个默认的主程序，从而节省时间。具体步骤如下：

（1）创建项目时选择 Spring Initializer，如图 10-7 所示。

（2）输入项目基本信息，如图 10-8 所示。

（3）接着选择需要添加的依赖，如图 10-9 所示。

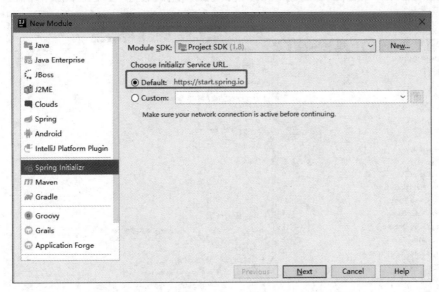

图 10-7　选择 Spring Initializer 构建 Spring Boot 项目

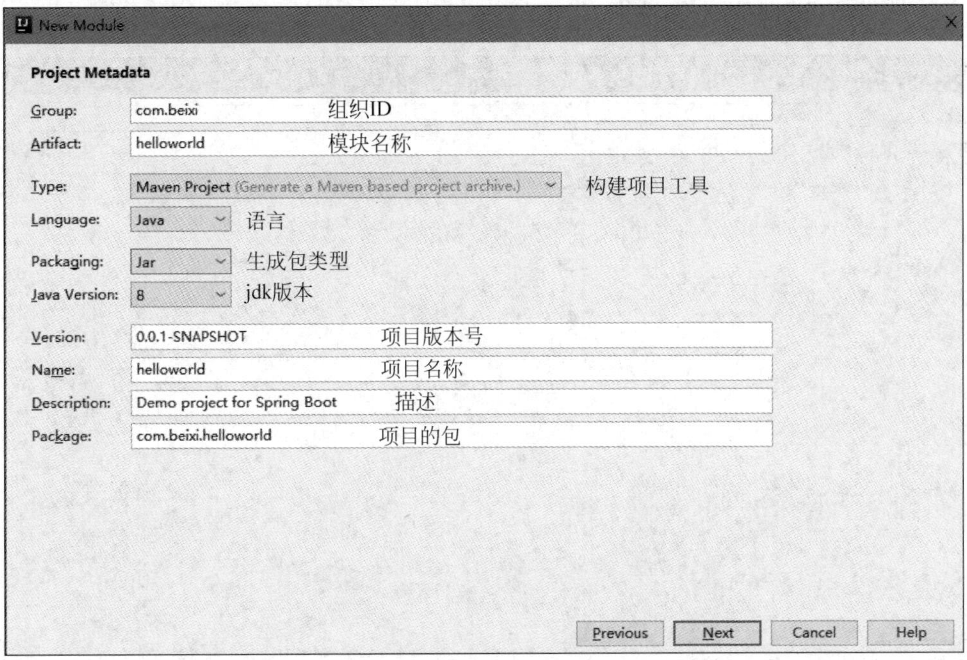

图 10-8　基本信息填写

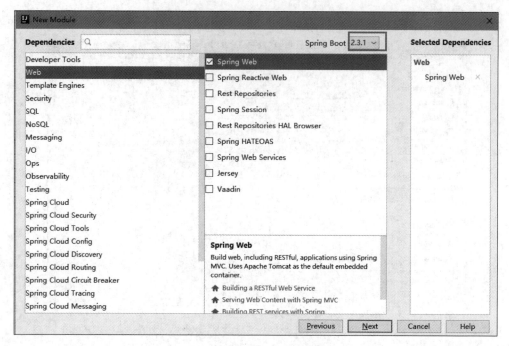

图 10-9 选择添加的依赖

(4) 选择项目创建的路径,如图 10-10 所示。

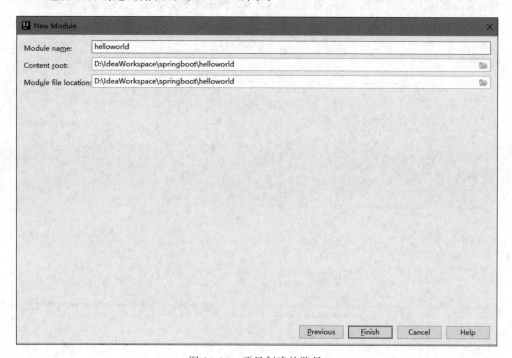

图 10-10 项目创建的路径

至此，Spring Boot 项目创建完成。

10.4 Spring Boot 项目结构介绍

10.4.1 目录结构

Spring Boot 的工程目录如图 10-11 所示。

（1）src/main/java：放置所有的 Java 源代码文件。

（2）src/main/resources/static：用于存放静态资源，如图片、CSS、JavaScript 等。

（3）src/main/resources/templates：用于存放 Web 页面的模板文件，如 Thymeleaf。

（4）src/main/resources/application.properties：配置信息，例如应用名、服务端口、数据库配置等，Spring Boot 默认支持两种配置文件类型，即.properties 和.yml。

（5）src/test：单元测试目录，生成的 ApplicationTests 通过 JUnit4 实现，可以直接运行 Spring Boot 应用的测试。

（6）target：放置编译后的.class 文件、配置文件等。

图 10-11　Spring Boot 的工程目录

10.4.2 启动类

Spring Boot 的启动类的类名是根据项目名称生成的，它的作用是启动 Spring Boot 项目，是基于 main 方法运行的。

启动类中@SpringBootApplication 注解：是一个组合注解，包含@SpringBootConfiguration＋@EnableAutoConfiguration＋@ComponentScan 3 个注解，是项目启动注解。

（1）@SpringBootConfiguration：是@Configuration 注解的派生注解，跟@Configuration 注解的功能一致，标注这个类是一个配置类，只不过@SpringBootConfiguration 是 springboot 的注解，而@Configuration 是 spring 的注解。

（2）@Configuration：通过对 Bean 对象的操作替代 spring 中 xml 文件。

（3）@EnableAutoConfiguration：Spring Boot 自动配置（auto-configuration），尝试根据添加的 jar 依赖自动配置 Spring 应用。是@AutoConfigurationPackage 和@Import (AutoConfigurationImportSelector.class) 注解的组合。

（4）@AutoConfigurationPackage：自动注入主类下所在包下所有的加了注解的类（@Controller、@Service 等），以及配置类（@Configuration）。

（5）@Import：直接导入普通的类。

（6）@ComponentScan：组件扫描，可自动发现和装配一些 Bean。

（7）@ConfigurationPropertiesScan：@ConfigurationPropertiesScan 扫描配置属性。

@EnableConfigurationProperties 注解的作用是使 @ConfigurationProperties 注解的类生效。

注意：启动类在启动时会做注解扫描（@Controller、@Service、@Repository…），扫描位置为同包或者子包下的注解，所以启动类的位置应放于包的根下。

10.4.3 POM 文件

1. 继承

Spring Boot 的父级依赖，只有继承如下依赖项目才是真正的 Spring Boot 项目，依赖如下：

```xml
<parent>
    <groupId>org.springframework.boot</groupId>
    <artifactId>spring-boot-starter-parent</artifactId>
    <version>2.3.1.RELEASE</version>
    <relativePath/><!-- lookup parent from repository -->
</parent>
```

spring-boot-starter-parent 是一个特殊的 starter，它用来提供相关的 Maven 默认依赖。使用它之后，常用的包依赖可以省去 version 标签。

2. 依赖

使用 Spring Initializer 快速创建的 Spring Boot 项目默认添加了如下依赖：

```xml
<dependency>
    <groupId>org.springframework.boot</groupId>
    <artifactId>spring-boot-starter-web</artifactId>
</dependency>
<dependency>
    <groupId>org.springframework.boot</groupId>
    <artifactId>spring-boot-starter-test</artifactId>
    <scope>test</scope>
</dependency>
```

（1）spring-boot-starter-web：spring-boot-starter-web 启动器支持全栈式 Web 开发，包括 Tomcat 和 spring-webmvc 等，引入该启动器，便能获得 Web 服务相关场景的 Jar 包。

（2）spring-boot-starter-test：引入这个启动器，就会把所有与测试相关场景的 Jar 包引入。

Spring Boot 提供了非常多的 starters 启动器。starters 启动器包含了一系列可以集成到应用里的依赖包，可以一站式集成 Spring 及其他技术，而不需要到处找示例代码和依赖包。例如如果要使用 Spring JPA 访问数据库，只要加入 spring-boot-starter-data-jpa 启动器依赖就能使用了。

starters 包含了许多项目中需要用到的依赖，它们能快速持续地运行，都是一系列得到支持的管理传递性依赖。

Spring Boot 官方的启动器都是以 spring-boot-starter-命名的,代表了一个特定的应用类型。

第三方的启动器不能以 spring-boot 开头命名,它们都被 Spring Boot 官方保留。一般一个第三方的启动器应该这样命名如:MyBatis 的 mybatis-spring-boot-starter。

3. 插件

spring-boot-maven-plugin 插件是将 Spring Boot 应用程序打包成 Jar 包的插件,代码如下:

```
<build>
  <plugins>
    <plugin>
      <groupId>org.springframework.boot</groupId>
      <artifactId>spring-boot-maven-plugin</artifactId>
    </plugin>
  </plugins>
</build>
```

将所有应用启动运行所需要的 Jar 包都包含进来,从逻辑上讲具备了独立运行的条件。可以使用"mvn package"进行打包,然后使用"java -jar 项目"命令就可以直接运行。

10.4.4　配置文件

配置文件的作用是修改 Spring Boot 自动配置的默认值。Spring Boot 提供一个名称为 application 的全局配置文件,支持两种格式 properteis 格式与 YAML 格式。

application.properties 文件是键值对类型的文件,之前一直在使用,所以不再对 properties 文件的格式进行阐述。

除了 properties 文件外,SpringBoot 还可以使用 yml 文件进行配置。YML 文件格式是 YAML(YAML Aint Markup Language)编写的文件格式,YAML 是一种直观的能够被计算机识别的数据序列化格式,并且容易被人类阅读,容易和脚本语言交互,可以被支持 YAML 库的不同编程语言程序导入,例如:C/C++、Ruby、Python、Java、Perl、C♯、PHP 等。YML 文件是以数据为核心的,是 JSON 的超集,简洁而强大,是一种专门用来书写配置文件的语言,可以代替 application.properties。

YAML 格式配置文件的扩展名可以是 yaml 也可以是 yml。

1. YAML 基本语法

(1) K:(空格)V 标识一对键值对。

(2) 以空格的缩进控制层级关系。

(3) 只要是左对齐的一列数据,都是同一层级的。

(4) 属性和值也对大小写敏感。

配置 Tomcat 监听端口,代码如下:

```
server:
    port: 8089
```

等同于 application.properties 中 Tomcat 监听端口的如下配置：

```
server.port = 8089
```

注意：冒号之后如果有值，那么冒号和值之间至少有一个空格，不能紧贴着。

2. 配置文件存放位置
（1）当前项目根目录中。
（2）当前项目根目录下的一个/config 子目录中。
（3）项目的 resources 即 classpath 根路径中。
（4）项目的 resources 即 classpath 根路径下的/config 目录中。

3. 配置文件加载顺序
（1）不同格式的加载顺序。

如果同一个目录下，既有 application.yml 又有 application.properties 文件，默认先读取 application.properties。

如果同一个配置属性，在多个配置文件都配置了，则默认使用第 1 个读取到的数据，后面读取的数据不覆盖前面读取到的数据。

（2）不同位置的加载顺序。
- 当前项目根目录下的一个/config 子目录中（最高）
    - config/application.properties
    - config/application.yml
- 当前项目根目录中（其次）
    - application.properties
    - application.yml

（3）项目的 resources 即 classpath 根路径下的/config 目录中（一般）。

```
resources/config/application.properties
resources/config/application.yml
```

（4）项目的 resources 即 classpath 根路径中（最后）。

```
resources/application.properties
resources/application.yml
```

10.5　Spring Boot 在 Controller 中的常用注解

1. @RestController
@RestController 注解包含了原来的@Controller 和@ResponseBody 注解，我们对

@Controller 注解已经非常了解了，这里不再赘述。@ResponseBody 注解是将返回的数据结构转换为 JSON 格式。所以 @RestController 可以看作 @Controller 和 @ResponseBody 的结合体，相当于偷了个懒，我们使用 @RestController 之后就不用再使用 @Controller 了。但是需要注意，如果是前后端分离，则不用模板渲染，例如 Thymeleaf，这种情况下是可以直接使用@RestController 将数据以 JSON 格式传给前端，前端得到之后解析。但如果不是前后端分离，则需要使用模板进行渲染，一般 Controller 都会返回到具体的页面，那么此时就不能使用@RestController 了，代码如下：

```
public String login() {
    return "login";
}
```

要想返回 login.html 页面，如果使用@RestController，会将 login 作为 JSON 字符串返回，所以这时候需要使用@Controller 注解。这在第 11 章 Spring Boot 集成 Thymeleaf 模板引擎中会再说明。

2．@GetMapping

@GetMapping 注解是@RequestMapping(method = RequestMethod.GET)的缩写。请求资源应该使用 GET。

3．@PostMapping

@PostMapping 注解是@RequestMapping(method = RequestMethod.POST)的缩写。添加资源应该使用 POST。

4．@PutMapping

@PutMapping 注解是@RequestMapping(method = RequestMethod.PUT)的缩写。更新资源应该使用 PUT。

5．@DeleteMapping

@DeleteMapping 注解是@RequestMapping(method = RequestMethod.DELETE)的缩写。删除资源应该使用 DELETE。

第 11 章 Spring Boot 整合 Web 开发

针对 Spring Boot 框架在 Web 应用中的开发使用问题,我们将探讨 Spring 框架的技术架构及模块组件,同时将讨论 Spring Boot 对于异常的处理,最后介绍 Spring Boot 如何使用过滤器、监听器和拦截器。

11.1 Spring Boot 访问静态资源

在 Spring Boot 项目中没有我们之前常规 Web 开发的 WebContent(WebApp),它只有 src 目录。在 src/main/resources 下面有两个文件夹 static 和 templates。

1) static

static 目录用于存放静态页面。Spring Boot 通过 classpath/static(classpath 指 resources 根目录)目录访问静态资源。

2) templates

templates 目录用于存放动态页面,在 Spring Boot 中不推荐使用 JSP 作为视图层技术,而是默认使用 Thymeleaf 来做动态页面。templates 目录则用于存放类似于 Thymeleaf 这样的模板引擎。

静态资源存放的其他位置如下。

1) Spring Boot 默认指定的可以存放静态资源的目录位置

(1) classpath:/META-INF/resources/ ##需创建/META-INF/resources/ 目录。

(2) classpath:/resources/ ##需创建/resources/目录。

(3) classpath:/static/ ##工具自动生成的 static 目录,也是用得最多的目录。

(4) classpath:/public/ ##需创建/public/目录。

在上面 4 个目录下存放静态资源(例如 login.html 等),可以直接访问(http://localhost:8080/login.html)。它们的优先级从上到下。所以,如果 static 里面有个 index.html,并且 public 下面也有个 index.html,则优先加载 static 下面的 index.html,因为优先级高,而且 Spring Boot 默认的首页是放在任一静态资源目录下的 index.html。

将静态资源放在上面指定的目录中,即可访问 index.html,如图 11-1 所示。Spring

Boot默认的Web页面图标可以放在任意静态资源目录下,访问时会自动在这4个目录下寻找,代码如下:

```html
//resources/public/index.html
<!DOCTYPE html>
<html lang = "en">
<head>
<meta charset = "UTF - 8">
<title>Title</title>
</head>
<body>
    <h4>Spring Boot 访问 Web 中的静态资源</h4>
    <img src = "favicon.jpg" width = "300px" height = "250px">
</body>
</html>
```

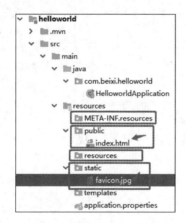

图 11-1　静态资源存放位置

2) 把静态资源打包成Jar包引入系统后供访问

由于最后会把Web项目打包成Jar包,并且发布到线上,引入Bootstrap、jQuery等静态资源文件就不能放在WebApp文件夹下(也没有WebApp文件夹),必须把静态资源打包成Jar包,添加至pom.xml文件,依赖查找移步WebJars官网(http://www.webjars.org/),使用Jar包的方式引入静态资源,类似于Maven仓库。

例如将jQuery引入到项目中,依赖如下:

```xml
<dependency>
    <groupId>org.webjars</groupId>
    <artifactId>jquery</artifactId>
    <version>3.3.1</version>
</dependency>
```

导入后,查看 org.webjars:jquery 的目录文件:

```
org.webjars:jquery:3.3.1
    -> jquery-3.3.1.jar
    -> META-INF
    -> maven
    -> resources
    -> webjars
        -> jquery
        -> 3.3.1
        -> jquery.js
        -> jquery.min.js
```

所有/webjars/*都从 classpath:/META-INF/resources/webjars/路径找对应的静态资源。

所以我们启动项目,访问:http://localhost:8080/webjars/jquery/3.3.1/jquery.js 即可。

11.2 整合 Thymeleaf

前端常用的模板引擎有 JSP、Velocity、Freemarker、Thymeleaf 等,模板引擎的作用是把数据和静态模板进行绑定,生成我们想要的 HTML。Spring Boot 推荐使用 Thymeleaf,因其语法简单、功能强大。

传统的 JSP+JSTL 组合已经过时了,Thymeleaf 是现代服务端的模板引擎,适用于 Web 和独立环境的现代服务器端 Java 模板引擎,与传统的 JSP 不同,Thymeleaf 可以使用浏览器直接打开,因为可以忽略拓展属性,相当于打开原生页面,给前端人员也带来一定的便利。Thymeleaf 的主要目标是将优雅的自然模板带到开发工作流程中,并将 HTML 在浏览器中正确显示,并且可以作为静态原型,让开发团队能更容易地协作。Thymeleaf 能够处理 HTML、XML、JavaScript、CSS 甚至纯文本。

11.2.1 Thymeleaf 使用

1. 引入 Thymeleaf

在 pom.xml 文件中引入 Thymeleaf 依赖,代码如下:

```xml
<dependency>
    <groupId>org.springframework.boot</groupId>
    <artifactId>spring-boot-starter-thymeleaf</artifactId>
</dependency>
```

默认版本为 Thymeleaf 2,如果想使用 Thymeleaf 3,则需添加如下配置:

```
<properties>
    <thymeleaf.version>3.0.2.RELEASE</thymeleaf.version>
    <thymeleaf-layout-dialect.version>2.1.1</thymeleaf-layout-
        dialect.version>
</properties>
```

2. Thymeleaf 使用

Spring Boot 针对 Thymeleaf 有一整套的自动化配置方案，这一套配置类的属性配置在 org.springframework.boot.autoconfigure.thymeleaf.ThymeleafProperties，部分源代码如下：

```
@ConfigurationProperties(prefix = "spring.thymeleaf")
public class ThymeleafProperties {
private static final Charset DEFAULT_ENCODING = Charset.forName("UTF-8");
private static final MimeType DEFAULT_CONTENT_TYPE = MimeType.valueOf("text/html");
public static final String DEFAULT_PREFIX = "classpath:/templates/";
public static final String DEFAULT_SUFFIX = ".html";
// ...
```

从源代码可以看出 Thymeleaf 模板的默认位置放在 classpath:/templates/目录下，默认的后缀是 html，这样 Thymeleaf 就能自动渲染。Thymeleaf 使用步骤如下：

在页面导入 Thymeleaf 的命名空间，以获得更好的提示，代码如下：

```
<html xmlns:th="http://www.thymeleaf.org">
```

创建 controller，代码如下：

```
//UserController
package com.beixi.helloworld.controller;
@Controller
public class UserController {
    @GetMapping("/index")
    public String index(Model model){
        List<User> list = new ArrayList<>();
        for (int i = 0; i < 5; i++) {
            User u = new User();
            u.setId(i);
            u.setName("贝西" + i);
            u.setAddress("山西" + i);
            list.add(u);
        }
        model.addAttribute("list", list);
```

```
            return "index";
        }
    }
    class User{
        private int id;
        private String name;
        private String address;
        //省略 get/set 方法
    }
```

在 controller 中我们返回视图层和数据,此时需要在 classpath:/templates/目录下新建一个视图层名为 index.html 的 Thymeleaf 模板文件。

注意:@Controller 注解需配置模板才能实现页面的跳转。@RestController 返回的是 JSON 数据,并不是视图页面。

创建 Thymeleaf 模板,代码如下:

```html
//index.html
<!DOCTYPE html>
<!-- 引入命名空间 -->
<html xmlns:th="http://www.thymeleaf.org">
<head>
<meta charset="UTF-8">
<title>Title</title>
</head>
<body>
<table border="1" width="60%" align="center">
    <tr>
        <td>编号</td>
        <td>姓名</td>
        <td>地址</td>
    </tr>
    <tr th:each="user:${list}">
        <td th:text="${user.id}"></td>
        <td th:text="${user.name}"></td>
        <td th:text="${user.address}"></td>
    </tr>
</table>
</body>
</html>
```

在 Thymeleaf 中,通过 th:each 指令遍历集合,数据的展示通过 th:text 实现。配置完成后启动项目,访问 http://localhost:8080/index,这样就可以看到数据集合了。

另外,Thymeleaf 支持在 JS 中直接获取 Model 存储的变量,代码如下:

```
@Controller
public class UserController {
    @GetMapping("/index")
    public String index(Model model){
        model.addAttribute("name", "贝西");
        return "index";
    }
}
```

在页面模板中直接通过 JS 获取，代码如下：

```
<script th:inline="javascript">
    var name = [[${name}]]
    console.log(name);
</script>
```

11.2.2 语法规则

1. 常用表达式

Thymeleaf 中常用的表达式，如表 11-1 所示。

表 11-1 Thymeleaf 表达式

表 达 式	作 用	表 达 式	作 用
${...}	变量表达式	#{...}	消息文字表达式
*{...}	选择表达式	@{...}	链接 URL 表达式

2. 常用的标签操作

Thymeleaf 中常用的标签，如表 11-2 所示。

表 11-2 Thymeleaf 中常用的标签

标 签	功 能	例 子
th:value	给属性赋值	\<input th:value="${值}" />
th:style	设置样式	th:style="'display:'+@{(${sitrue}?'none':'inline-block')}+''"
th:onclick	单击事件	th:onclick="方法名(值)"
th:if	条件判断	\<div th:if="*{num==50}">等于 50\</div>
th:unless	条件判断和 th:if 相反	\<a th:href="@{/user/login}" th:unless=${session.user != null}> Login \</a>
th:href	超链接	\<a th:href="@{/user/login}"> Login \</a>
th:switch	配合 th:case	\<div th:switch="${user.role}">
th:case	配合 th:switch	\<p th:case="'admin'"> administrator \</p>
th:src	图片类地址引入	\<img alt="csdn logo" th:src="@{/img/logo.png}" />
th:action	表单提交的地址	\<form th:action="@{/user/update}">

3. 常用函数

Thymeleaf 中常用的函数，如表 11-3 所示。

表 11-3　常用的函数

函　　数	作　　用	函　　数	作　　用
#dates	日期函数	#numbers	数字函数
#lists	列表函数	#calendars	日历函数
#arrays	数组函数	#objects	对象函数
#strings	字符串函数	#bools	逻辑函数

Thymeleaf 还有很多其他用法，具体用法可以参考 Thymeleaf 的官方文档（https://www.thymeleaf.org/doc/tutorials/3.0/usingthymeleaf.html）。这里主要学会如何在 Spring Boot 中使用 Thymeleaf，遇到对应的标签或者方法，查阅官方文档即可。

11.3　Spring Boot 返回 JSON 数据

在项目开发中，接口与接口之间，以及前后端之间数据的传输都使用 JSON 格式。在 Spring Boot 中，使接口返回 JSON 格式的数据很简单，在 Controller 中使用 @RestController 注解即可返回 JSON 格式的数据，@RestController 也是 Spring Boot 新增的一个注解，它包含了原来的 @Controller 和 @ResponseBody 注解，@ResponseBody 注解是将返回的数据结构转换为 JSON 格式。

在默认情况下，使用 @RestController 注解即可将返回的数据结构转换成 JSON 格式，在 Spring Boot 中默认使用的 JSON 解析技术框架是 Jackson。我们打开 pom.xml 文件中的 spring-boot-starter-web 依赖，可以看到 spring-boot-starter-json 依赖：

```
<dependency>
    <groupId>org.springframework.boot</groupId>
    <artifactId>spring-boot-starter-json</artifactId>
    <version>2.3.1.RELEASE</version>
    <scope>compile</scope>
</dependency>
```

Spring Boot 对依赖做了很好的封装，可以看到很多 spring-boot-starter-xxx 系列的依赖，这是 Spring Boot 的特点之一，不需要人为引入很多相关的依赖，starter-xxx 系列直接包含了必要的依赖，所以我们再次打开上面提到的 spring-boot-starter-json 依赖文件，可以看到如下代码：

```
<dependency>
    <groupId>com.fasterxml.jackson.core</groupId>
```

```xml
    <artifactId>jackson-databind</artifactId>
    <scope>compile</scope>
</dependency>
<dependency>
    <groupId>com.fasterxml.jackson.datatype</groupId>
    <artifactId>jackson-datatype-jdk8</artifactId>
    <scope>compile</scope>
</dependency>
<dependency>
    <groupId>com.fasterxml.jackson.datatype</groupId>
    <artifactId>jackson-datatype-jsr310</artifactId>
    <scope>compile</scope>
</dependency>
<dependency>
    <groupId>com.fasterxml.jackson.module</groupId>
    <artifactId>jackson-module-parameter-names</artifactId>
    <scope>compile</scope>
</dependency>
```

到此为止，我们知道了 Spring Boot 中默认使用的 JSON 解析框架是 Jackson。下面我们看一下默认的 Jackson 框架对常用数据类型的转 JSON 处理。

11.3.1　常用数据类型转为 JSON 格式

在实际项目中，常用的数据结构无非有类对象、List 对象、Map 对象，我们看一下默认的 Jackson 框架如何将这三种常用的数据结构转成 JSON 格式。

1. 创建实体类

创建 User 类，代码如下：

```java
public class User {
    private int id;
    private String name;
    private String password;
    /* 省略 get、set 和带参构造方法 */
}
```

2. 创建 Controller 类

接着创建一个 Controller 类，分别返回 User 对象、List 和 Map<String, Object>，代码如下：

```java
@RestController
@RequestMapping("/json")
```

```java
public class JsonController {
    @RequestMapping("/user")
    public User getUser() {
        return new User(10,"贝西","11");
    }
    @RequestMapping("/list")
    public List<User> getUserList() {
        List<User> userList = new ArrayList<>();
        User user1 = new User(1, "贝西", "123456");
        User user2 = new User(2, "贾志杰", "123456");
        userList.add(user1);
        userList.add(user2);
        return userList;
    }
    @RequestMapping("/map")
    public Map<String, Object> getMap() {
        Map<String, Object> map = new HashMap<>(3);
        User user = new User(1, "贾志杰", "123456");
        map.put("作者信息", user);
        map.put("博客地址", "https://blog.csdn.net/beixishuo");
        map.put("公众号","贝西奇谈");
        map.put("B站","贝西贝西");
        return map;
    }
}
```

3. 测试不同数据类型返回的 JSON 格式

控制层接口完成后,分别返回了 User 对象、List 集合和 Map 集合。接下来我们依次测试一下效果。

在浏览器中输入 localhost:8080/json/user,返回 JSON 格式,代码如下:

```
{id: 10,name: "贝西",password: "11"}
```

在浏览器中输入 localhost:8080/json/list,返回 JSON 格式,代码如下:

```
[{"id":1,"name":"贝西","password":"123456"},{"id":2,"name":"贾志杰","password":"123456"}]
```

在浏览器中输入 localhost:8080/json/map,返回 JSON 格式,代码如下:

```
{"作者信息":{"id":1,"name":"贾志杰","password":"123456"},"博客地址":"https://blog.csdn.net/beixishuo","公众号":"贝西奇谈","B站":"贝西贝西"}
```

11.3.2　Jackson 中对 null 的处理

在实际项目中，难免会遇到一些 null 值。当转 JSON 格式时，不希望这些 null 值出现，例如我们希望所有的 null 值在转 JSON 格式时都变成空字符串。

在 Spring Boot 中，我们做一下配置即可，新建一个 Jackson 的配置类：

```
@Configuration
public class JacksonConfig {
    @Bean
    @Primary
    @ConditionalOnMissingBean(ObjectMapper.class)
    public ObjectMapper jacksonObjectMapper(Jackson2ObjectMapperBuilder builder) {
        ObjectMapper objectMapper = builder.createXmlMapper(false).build();
        objectMapper.getSerializerProvider().setNullValueSerializer(new JsonSerializer<Object>() {
            @Override
            public void serialize(Object o, JsonGenerator jsonGenerator, SerializerProvider serializerProvider) throws IOException {
                jsonGenerator.writeString("");
            }
        });
        return objectMapper;
    }
}
```

然后修改一下上面返回的 Map 接口，将几个值改成 null 进行测试，代码如下：

```
@RequestMapping("/map")
public Map<String, Object> getMap() {
    Map<String, Object> map = new HashMap<>(3);
    User user = new User(1, "贾志杰", null);
    map.put("作者信息", user);
    map.put("博客地址", "https://blog.csdn.net/beixishuo");
    map.put("公众号", "贝西奇谈");
    map.put("B站", null);
    return map;
}
```

重启项目，再次输入 localhost:8080/json/map，可以看到 Jackson 已经将所有 null 字段转成空字符串了。

11.3.3　封装统一返回的数据结构

以上是 Spring Boot 返回 JSON 格式的几个有代表性的例子，但是在实际项目中，除了要封装数据之外,往往需要在返回的 JSON 格式中添加一些其他信息,例如返回一些状态码 code,返回一些 msg 给调用者,这样调用者可以根据 code 或者 msg 做一些逻辑判断。所以在实际项目中,我们需要封装一个统一的 JSON 返回结构用于存储返回信息。

1. 定义统一 JSON 结构

由于封装的 JSON 数据的类型不确定,所以在定义统一的 JSON 结构时,我们需要用到泛型。统一的 JSON 结构中属性包括数据、状态码、提示信息即可,构造方法可以根据实际业务需求做相应的添加,一般来说,应该有默认的返回结构,也应该有用户指定的返回结构,代码如下:

```java
public class JsonResult<T> {

    private T data;
    private String code;
    private String msg;
    /**
     * 若没有数据返回,默认状态码为 0,提示信息为:操作成功!
     */
    public JsonResult() {
        this.code = "0";
        this.msg = "操作成功!";
    }
    /**
     * 若没有数据返回,可以人为指定状态码和提示信息
     * @param code
     * @param msg
     */
    public JsonResult(String code, String msg) {
        this.code = code;
        this.msg = msg;
    }
    /**
     * 有数据返回时,状态码为 0,默认提示信息为:操作成功!
     * @param data
     */
    public JsonResult(T data) {
        this.data = data;
        this.code = "0";
        this.msg = "操作成功!";
    }
    /**
     * 有数据返回,状态码为 0,人为指定提示信息
     * @param data
     * @param msg
```

```
     */
    public JsonResult(T data, String msg) {
        this.data = data;
        this.code = "0";
        this.msg = msg;
    }
    // 省略 get 和 set 方法
}
```

2. 修改 Controller 中的返回值类型及测试

由于 JsonResult 使用了泛型，所以所有的返回值类型都可以使用该统一结构，在具体的场景将泛型替换成具体的数据类型即可，非常方便，也便于维护。根据以上的 JsonResult，我们改写一下 Controller，代码如下：

```
@RestController
@RequestMapping("/jsonresult")
public class JsonController {
    @RequestMapping("/user")
    public JsonResult<User> getUser() {
        User user = new User(10, "贝西", "11");
        return new JsonResult<>(user);
    }

    @RequestMapping("/list")
    public JsonResult<List<User>> getUserList() {
        List<User> userList = new ArrayList<>();
        User user1 = new User(1, "贝西", "123456");
        User user2 = new User(2, "贾志杰", "123456");
        userList.add(user1);
        userList.add(user2);
        return new JsonResult<>(userList, "获取用户列表成功");
    }
    @RequestMapping("/map")
    public JsonResult<Map<String, Object>> getMap() {
        Map<String, Object> map = new HashMap<>(3);
        User user = new User(1, "贾志杰", "123456");
        map.put("作者信息", user);
        map.put("博客地址", "https://blog.csdn.net/beixishuo");
        map.put("公众号", "贝西奇谈");
        map.put("B站", "贝西贝西");
        return new JsonResult<>(map);
    }
}
```

可以重新在浏览器中输入 localhost:8080/jsonresult/user，返回 JSON 格式，代码

如下：

```
{"code":"0","data":{"id":10,"name":"贝西","password":"11"},"msg":"操作成功!"}
```

在浏览器中输入 localhost:8080/jsonresult/list，返回 JSON 格式，代码如下：

```
{"code":"0","data":[{"id":1,"name":"贝西","password":"123456"},{"id":2,"name":"贾志杰","password":"123456"}],"msg":"获取用户列表成功"}
```

在浏览器中输入 localhost:8080/jsonresult/map，返回 JSON 格式，代码如下：

```
{"code":"0","data":{"作者信息":{"id":1,"name":"贾志杰","password":"123456"},"博客地址":"https://blog.csdn.net/beixishuo","公众号":"贝西奇谈"},"msg":"操作成功!"}
```

通过封装，不但将数据通过 JSON 传给前端或者其他接口，还附上了状态码和提示信息，这在实际项目场景中用得非常广泛。

11.4　Spring Boot 中的异常处理

在项目开发过程中，不管是对底层数据库的操作，还是业务层的处理，以及控制层的处理，都不可避免地会遇到各种可预知的、不可预知的异常需要处理。Spring Boot 框架异常处理有 5 种处理方式，从范围来说包括全局异常捕获处理方式和局部异常捕获处理方式，接下来通过使用除数不能为 0 的后端异常代码，讲解这 5 种捕获方式，代码如下：

```java
//后端异常代码
@Controller
public class ExceptionController {
    private static final Logger log = LoggerFactory.getLogger(ExceptionController.class);

    @RequestMapping("/exceptionMethod")
    public String exceptionMethod(Model model) throws Exception {

        model.addAttribute("msg", "没有抛出异常");
        /* 抛出 java.lang.ArithmeticException: / by zero 异常 */
        int num = 1/0;
log.info(String.valueOf(num));
        return "index";
    }
}
```

11.4.1　自定义异常错误页面

在遇到异常时，Spring Boot 会自动跳转到一个默认的异常页面，如请求上述 http://

localhost:8080/exceptionMethod 路径时发生状态值为 500 的错误，Spring Boot 会有一个默认的页面展示给用户，如图 11-2 所示。

图 11-2　发生状态值为 500 的默认异常展示页面

事实上，Spring Boot 在返回错误信息时不一定返回 HTML 页面，而是根据实际情况返回 HTML 或者 JSON 数据（若开发者发起 Ajax 请求，则错误信息返回 JSON 数据）。

Spring Boot 默认的异常处理机制是程序中出现了异常 Spring Boot 就会请求/error 的 URL。在 Spring Boot 中的错误默认是由 BasicExceptionController 类来处理/error 请求，然后跳转到默认显示异常的页面来展示异常信息。介绍到这里，读者已经可以发现要自定义异常错误页面 error 了，这里以 Thymeleaf 为例，Thymeleaf 页面模板默认处于 classpath:/templates/下，因此在该目录下创建 error.html 文件即可，代码如下：

```
//error.html
<!DOCTYPE html>
<html xmlns:th="http://www.thymeleaf.org">
<head>
<meta charset="UTF-8">
<title>自定义 springboot 异常处理页面</title>
</head>
<body>
    Springboot BasicExceptionController 错误页面
    <br>
    <span th:text=""${msg}"></span>
</body>
</html>
```

在指定目录添加 error.html 页面后再访问/exceptionMethod 接口，则页面效果如图 11-3 所示。

图 11-3　添加 error.html 页面后的页面效果

11.4.2 使用@ExceptionHandler注解处理局部异常

Spring MVC提供了@ExceptionHandler这个注解，在Spring Boot里面，我们同样可以使用它来做异常捕获。直接在对应的Controller里增加一个异常处理的方法，并使用@ExceptionHandler标识它即可，属于局部处理异常，代码如下：

```
@Controller
public class ExceptionController {
    private static final Logger log = LoggerFactory.getLogger(ExceptionController.class);
    @RequestMapping("/exceptionMethod")
    public String exceptionMethod(Model model) throws Exception {
        model.addAttribute("msg", "没有抛出异常");
        int num = 1/0;
        log.info(String.valueOf(num));
        return "index";
    }
    /**
     * 描述：捕获ExceptionController中的ArithmeticException异常
     * @param model 将Model对象注入方法中
     * @param e 将产生异常对象注入方法中
     * @return 指定错误页面
     */
    @ExceptionHandler(value = {ArithmeticException.class})
    public String arithmeticExceptionHandle(Model model, Exception e) {
        model.addAttribute("msg","@ExceptionHandler" + e.getMessage()); log.info(e.getMessage());
        return "error";
    }
}
```

@ExceptionHandler拦截了异常，我们可以通过该注解实现自定义异常处理。其中，@ExceptionHandler配置的value指定需要拦截的异常类型。

这样只能做到单一的controller异常处理，项目中一般都存在着多个Controller，它们对于大多数异常处理的方法都大同小异，这样就合适在每一个Controller里都编写一个对应的异常处理方法，所以不推荐使用。

当访问http://localhost:8080/exceptionMethod时，跳转到的页面如图11-4所示，显示@ExceptionHandler/by zero，控制台不再报错，表示我们使用@ExceptionHandler注解处理异常成功。

图11-4 @ExceptionHandler注解处理异常

11.4.3　使用 @ControllerAdvice 注解处理全局异常

实际开发中，需要对异常分门别类地进行处理，使用@ControllerAdvice＋@ExceptionHandler 注解能够处理全局异常，推荐使用这种方式，可以根据不同的异常对不同的异常进行处理。

使用方式：定义一个类，使用 @ControllerAdvice 注解该类，使用 @ExceptionHandler 注解方法，这里笔者定义了一个 GlobalException 类表示用来处理全局异常，代码如下：

```java
//GlobalException.java
@ControllerAdvice
public class GlobalException {
    private static final Logger log = LoggerFactory.getLogger(GlobalException.class);
    /**
     * 描述：捕获 ArithmeticException 异常
     * @param model 将 Model 对象注入方法中
     * @param e 将产生异常对象注入方法中
     * @return 指定错误页面
     */
    @ExceptionHandler(value = {ArithmeticException.class})
    public String arithmeticExceptionHandle(Model model, Exception e) {
        model.addAttribute("msg", "@ControllerAdvice + @ExceptionHandler :" + e.getMessage());
        log.info(e.getMessage());
        return "error";
    }
}
```

如果需要处理其他异常，如 NullPointerException 异常，则只需在 GlobalException 类中定义一个方法并使用 @ExceptionHandler(value＝{NullPointerException.class})注解该方法，在该方法内部处理异常就可以了。

当访问/exceptionMethod 接口时，页面显示效果如图 11-5 所示，显示@ControllerAdvice＋@ExceptionHandler:/by zero，表示我们使用@ControllerAdvice＋@ExceptionHandler 注解全局处理异常成功。

图 11-5　@ControllerAdvice 处理异常

11.4.4　配置 SimpleMappingExceptionResolver 类处理异常

通过配置 SimpleMappingExceptionResolver 类处理异常也是全局范围的，通过将 SimpleMappingExceptionResolver 类注入 Spring 容器处理异常，代码如下：

```java
@Configuration
public class GlobalException {
```

```
@Bean
public SimpleMappingExceptionResolver
getSimpleMappingExceptionResolver(){
    SimpleMappingExceptionResolver resolver = new SimpleMappingExceptionResolver();

    Properties mappings = new Properties();
    /*
     * 参数一：异常的类型,注意必须是异常类型的全名
     * 参数二：视图名称
     */
    mappings.put("java.lang.ArithmeticException", "error");
    //设置异常与视图映射信息
    resolver.setExceptionMappings(mappings);
    return resolver;
}
```

注意：在类上添加@Configuration 注解，在方法上添加@Bean 注解，方法返回值必须是 SimpleMappingExceptionResolver。

代码说明如下。

（1）用@Bean 注解的方法：会实例化、配置并初始化一个新的对象，这个对象会由 Spring IoC 容器管理。

（2）@Configuration：从定义看，用于注解类、接口、枚举、注解的定义。@Configuration 用于类，表明这个类是 beans 定义的类。

访问/exceptionMethod 接口后抛出 ArithmeticException 异常，跳转到 error.html 页面，效果如图 11-6 所示。

图 11-6　SimpleMappingExceptionResolver 类处理异常

11.4.5　实现 HandlerExceptionResolver 接口处理异常

通过实现 HandlerExceptionResolver 接口处理异常，首先编写类实现 HandlerExceptionResolver 接口，代码如下：

```
@Configuration
public class HandlerExceptionResolverImpl implements HandlerExceptionResolver {
    @Override
    public ModelAndView resolveException(HttpServletRequest request, HttpServletResponse response, Object handler,Exception ex) {
        ModelAndView modelAndView = new ModelAndView();
        modelAndView.addObject("msg", "实现 HandlerExceptionResolver 接口处理异常");
```

```
        //判断不同异常类型,实现不同视图跳转
        if(ex instanceof ArithmeticException){
            modelAndView.setViewName("error");
        }
        return modelAndView;
    }
}
```

配置完后访问/exceptionMethod 接口,效果如图 11-7 所示。

至此 SpringBoot 框架异常处理的 5 种方式就全部讲解完了。

11.4.6 一劳永逸

图 11-7 实现 HandlerExceptionResolver 接口处理异常

当然了,异常很多,例如还有 RuntimeException,数据库还有一些查询或者操作异常等。由于 Exception 异常是父类,所有异常都会继承该异常,所以我们可以直接拦截 Exception 异常,一劳永逸,代码如下:

```
@ControllerAdvice
public class GlobalException{
    private static final Logger log = LoggerFactory.getLogger(GlobalException.class);
    /**
     * 系统异常,预期以外异常
     * @param e
     * @return
     */
    @ExceptionHandler(Exception.class)
//@ResponseStatus(value = HttpStatus.INTERNAL_SERVER_ERROR)
    public JsonResult handleUnexpectedServer(Model model,Exception ex) {
        model.addAttribute("msg", "系统发生异常,请联系管理员");
        log.info(e.getMessage());
        return "error";
    }
}
```

访问/exceptionMethod 接口,效果如图 11-8 所示。

图 11-8　一劳永逸

在项目中,我们一般都会比较详细地去拦截一些常见异常,拦截 Exception 虽然可以一劳永逸,但是不利于我们去排查或者定位问题。实际项目中,可以把拦截 Exception 异常写在 GlobalException 最下面,如果还是没有找到常见异常,最后再拦截一下 Exception 异常,保证异常得到处理。

11.5 配置嵌入式 Servlet 容器

没有使用 Spring Boot 开发时,需要安装 Tomcat 环境,项目打包成 War 包后进行部署。而 Spring Boot 默认使用 Tomcat 作为嵌入式的 Servlet 容器。

11.5.1 如何定制和修改 Servlet 容器的相关配置

在内置的 Tomcat 中,不再有 web.xml 文件可以供我们修改,在 Spring Boot 中修改 Servlet 容器相关的配置有两种方式可供选择,一种是在配置文件中修改,另一种是通过配置类的方式去修改。

(1) 在配置文件中修改(具体修改的参数可以查看 ServerProperties 类),代码如下:

```
spring.mvc.date-format = yyyy-MM-dd
spring.thymeleaf.cache = false
spring.messages.basename = i18n.login
server.port = 8081
server.context-path = /
server.tomcat.uri-encoding = UTF-8
```

只需在 application.properties 或者 application.yml/yaml 中像上面那样就可以轻松地修改相关的配置。

(2) 编写一个 WebServerFactoryCustomizer:嵌入式的 Servlet 容器定制器,来修改 Servlet 容器的配置。

新建 MyMvcConfig 类,代码如下:

```
@Configuration
public class MyMvcConfig {
    @Bean
    public WebServerFactoryCustomizer<ConfigurableWebServerFactory> webServerFactoryCustomizer(){
        return new WebServerFactoryCustomizer<ConfigurableWebServerFactory>() {
            @Override
            public void customize(ConfigurableWebServerFactory factory) {
                factory.setPort(8081);
            }
        };
    }
}
```

11.5.2 注册Servlet三大组件——Servlet、Filter、Listener

一般情况下,使用Spring、Spring MVC等框架后,几乎不需要再使用Servlet、Filter、Listener了,但有时在整合一些第三方框架时,可能还是不得不使用Servlet,如在整合报表插件时就需要使用Servlet。Spring Boot对整合这些基本的Web组件也提供了很好的支持。

由于Spring Boot默认是以Jar包的方式启动嵌入式的Servlet容器从而启动Spring Boot的Web应用,没有使用web.xml文件。所以可以用下面的方式在Spring Boot项目中添加3个组件:

```java
@WebServlet("/servlet")
public class MyServlet extends HttpServlet {
    @Override
    protected void doGet(HttpServletRequest req, HttpServletResponse resp) throws ServletException, IOException {
        doPost(req, resp);
    }
    @Override
    protected void doPost(HttpServletRequest req, HttpServletResponse resp) throws ServletException, IOException {
        resp.getWriter().write("Hello MyServlet");
        System.out.println("name:" + req.getParameter("name"));
    }
}


@WebFilter("/")
public class MyFilter implements Filter {
    @Override
    public void init(FilterConfig filterConfig) throws ServletException {
        System.out.println("MyFilter -- init");
    }
    @Override
    public void doFilter(ServletRequest servletRequest, ServletResponse servletResponse, FilterChain filterChain) throws IOException, ServletException {
        System.out.println("myFilter -- doFilter");
        filterChain.doFilter(servletRequest, servletResponse);
    }
    @Override
    public void destroy() {
        System.out.println("MyFilter -- destroy");
    }
}
```

```java
}

@WebListener
public class MyListener implements ServletContextListener {
    @Override
    public void contextInitialized(ServletContextEvent servletContextEvent) {
        System.out.println("Web项目启动了…");
    }
    @Override
    public void contextDestroyed(ServletContextEvent servletContextEvent) {
        System.out.println("Web项目销毁了…");
    }
}
```

当然想要使用三大组件的注解,就必须先在 Spring Boot 主配置类(即标注了@SpringBootApplication 注解的类)上添加 @ServletComponentScan 注解,以实现对 Servlet、Filter 及 Listener 的扫描,代码如下:

```java
@ServletComponentScan
@SpringBootApplication
public class HelloworldApplication {
    public static void main(String[] args) {
        SpringApplication.run(HelloworldApplication.class, args);
    }
}
```

启动项目,在浏览器中输入 http://localhost:8080/servlet? name=beixi,在控制台查看日志信息,如图 11-9 所示。

```
2020-08-15 00:48:25.446  INFO 2372
web项目启动了。。。
MyFilter--init
2020-08-15 00:48:25.691  INFO 2372
2020-08-15 00:48:25.766  INFO 2372
2020-08-15 00:48:25.907  INFO 2372
2020-08-15 00:48:25.918  INFO 2372
2020-08-15 00:48:29.441  INFO 2372
 Note: further occurrences of this
name:beixi
```

图 11-9 三大组件注册信息

11.5.3 替换为其他嵌入式 Servlet 容器

Spring Boot 默认使用的是 Tomcat，当然也可以切换成其他的容器，而且切换的方式也很简单，只需引入其他容器的依赖将当前容器的依赖排除。

jetty 比较适合做长连接的项目，例如聊天等这种一直要连接网络进行通信的项目。

想要将容器从 Tomcat 切换到 jetty，可在 pom.xml 文件中导入相关依赖：

```xml
<dependency>
    <groupId>org.springframework.boot</groupId>
    <artifactId>spring-boot-starter-web</artifactId>
    <exclusions><!-- 移除 Tomcat -->
        <exclusion>
            <artifactId>spring-boot-starter-tomcat</artifactId>
            <groupId>org.springframework.boot</groupId>
        </exclusion>
    </exclusions>
</dependency>
<!-- 引入其他的 Servlet 容器 -->
<dependency>
    <groupId>org.springframework.boot</groupId>
    <artifactId>spring-boot-starter-jetty</artifactId>
</dependency>
```

undertow 不支持 JSP，但是它是一个高性能的非阻塞的 Servlet 容器，并发性能好。引入 undertow 的方式与 jetty 一样，添加依赖代码如下：

```xml
<dependency>
    <groupId>org.springframework.boot</groupId>
    <artifactId>spring-boot-starter-web</artifactId>
    <exclusions>
        <exclusion>
            <artifactId>spring-boot-starter-tomcat</artifactId>
            <groupId>org.springframework.boot</groupId>
        </exclusion>
    </exclusions>
</dependency>
<!-- 引入其他的 Servlet 容器 -->
<dependency>
    <groupId>org.springframework.boot</groupId>
    <artifactId>spring-boot-starter-undertow</artifactId>
</dependency>
```

11.6 在 Spring Boot 中使用拦截器

Spring Boot 延续了 Spring MVC 提供的 AOP 风格拦截器,拥有精细的拦截处理能力,在 Spring Boot 中拦截器的使用更加方便。这里只是用登录的例子来展现拦截器的基本使用。拦截器用途很广,例如可以对 URL 路径进行拦截,也可以用于权限验证、解决乱码、操作日志记录、性能监控、异常处理等。

在项目中创建 interceptor 包,并创建一个 LoginInterceptor 拦截器实现 HandlerInterceptor 接口。

一般用户登录功能我们可以这样实现:要么往 session 中写一个 user,要么针对每一个 user 生成一个 token。第二种方式更好,针对第二种方式,如果用户登录成功了,则每次请求的时候都会带上该用户的 token;如果未登录成功,则没有该 token,服务端可以通过检测这个 token 参数的有无来判断用户有没有登录成功,从而实现拦截功能。代码如下:

```java
/*自定义拦截器*/
public class LoginInterceptor implements HandlerInterceptor {
    private static final Logger logger = LoggerFactory.getLogger(LoginInterceptor.class);
    @Override
     public boolean preHandle(HttpServletRequest request, HttpServletResponse response, Object handler) throws Exception {
        HandlerMethod handlerMethod = (HandlerMethod) handler;
        Method method = handlerMethod.getMethod();
        String methodName = method.getName();
        logger.info("====拦截到了方法:{},在该方法执行之前执行====", methodName);

        // 判断用户有没有登录成功,一般登录成功的用户都有一个对应的 token
        String token = request.getParameter("token");
        if (null == token || "".equals(token)) {
            logger.info("用户未登录,没有权限执行……请登录");
            return false;
        }

        // 返回 true 时才会继续执行,返回 false 则取消当前请求
        return true;
    }
    @Override
    public void postHandle(HttpServletRequest request, HttpServletResponse response, Object handler, ModelAndView modelAndView) throws Exception {
        logger.info("执行完方法之后执行(Controller 方法调用之后),但是此时还没进行视图渲染");
    }
    @Override
```

```
        public void afterCompletion(HttpServletRequest request, HttpServletResponse response,
Object handler, Exception ex) throws Exception {
            logger.info("整个请求都已处理完,DispatcherServlet 也渲染了对应的视图,此时可以
做一些清理工作");
        }
    }
```

每一个拦截器都需要实现 HandlerInterceptor 接口,实现这个接口有 3 种方法,每种方法会在请求调用的不同时期完成,因为我们需要在接口调用之前拦截请求并判断是否登录成功,所以这里需要使用 preHandle 方法,在里面写验证逻辑,最后返回 true 或者 false,确定请求是否合法。

(1) 通过配置类注册拦截器。

创建一个配置类 InterceptorConfig,并实现 WebMvcConfigurer 接口,覆盖接口中的 addInterceptors 方法,并为该配置类添加 @Configuration 注解,标注此类为一个配置类,让 Spring Boot 扫描得到,代码如下:

```
@Configuration
public class InterceptorConfig implements WebMvcConfigurer {
    @Override
    public void addInterceptors(InterceptorRegistry registry) {
        //需要拦截的路径,/** 表示需要拦截所有请求
        String[] addPathPatterns = {"/**"};
        //不需要拦截的路径,表示除了登录与注册之外,因为注册不需要登录也可以访问
        String [] excludePathPaterns = {
                "/login.html",
                "/registry.html"
        };
        //注册一个登录拦截器
        registry.addInterceptor(new LoginInterceptor())
                .addPathPatterns(addPathPatterns)
                .excludePathPatterns(excludePathPaterns);
        //注册一个权限拦截器,如果有多个拦截器,只需添加以下一行代码
        //registry.addInterceptor(new LoginInterceptor())
        //  .addPathPatterns(addPathPatterns)
        //  .excludePathPatterns(excludePathPatterns);
    }
}
```

(2) 登录测试类。

创建 UserController,用于验证拦截器是否可行,代码如下:

```
@Controller
@RequestMapping("/interceptor")
```

```java
public class UserController {
    @RequestMapping("/test")
    public String test() {
        return "hello.html";
    }
}
```

让其跳转到 hello.html 页面,直接在 hello.html 中输出 hello interceptor 即可。

启动项目,在浏览器中输入 localhost:8080/interceptor/test 后查看控制台日志,发现请求被拦截,如图 11-10 所示。

```
-1] c.b.h.interceptor.LoginInterceptor    : ====拦截到了方法: test,在该方法执行之前执行====
-1] c.b.h.interceptor.LoginInterceptor    : 用户未登录,没有权限执行......请登录
```

图 11-10 拦截器拦截日志信息

如果在浏览器中输入:localhost:8080/interceptor/test?token=123 即可正常运行。

第 12 章 应用开发

12.1 文件上传与下载

Spring Boot 没有自己的文件上传与下载技术,它依赖于 Spring MVC 的文件上传与下载技术,只不过在 Spring Boot 中做了更进一步的简化,更为方便。本节将介绍在 Spring MVC 中实现文件上传与下载的技术,并通过示例展示该技术在 Spring Boot 项目中的应用。

12.1.1 单文件上传

上传文件必须将表单 method 设置为 POST,并将 enctype 设置为 multipart/form-data。只有这样,浏览器才会把用户所选文件的二进制数据发送给服务器。Spring MVC 在文件上传时,会将上传的文件映射为 MultipartFile 对象,并对 MultipartFile 对象进行文件的解析和保存。

MultipartFile 接口有以下几个常用方法。

(1) byte[ ] getBytes():获取文件数据。

(2) String getContentType():获取文件 MIME 类型,如 application/pdf、image/pdf 等。

(3) InputStream getInputStream():获取文件流。

(4) String getOriginalFileName():获取上传文件的原名称。

(5) long getSize():获取文件的字节大小,单位为 Byte。

(6) boolean isEmpty():是否有上传的文件。

(7) void transferTo(File dest):将上传的文件保存到一个目标文件中。

首先创建一个 Spring Boot 项目并添加 spring-boot-starter-web 依赖。然后在 src/main/resources 目录下的 static 目录下新建 upload.html 文件,代码如下:

```
//upload.html
<!DOCTYPE html>
<html>
```

```html
<head>
<meta http-equiv="Content-type" content="text/html;charset=UTF-8">
<meta http-equiv="X-UA-Compatible" content="IE=edge,chrome=1"/>
<title>单文件上传</title>
</head>
<body>
    <form method="post" action="/file/upload" enctype="multipart/form-data">
        <input type="file" name="file">
        <input type="submit" value="提交">
    </form>
</body>
</html>
```

接下来在 application.yml 文件中添加如下配置：

```yaml
server:
  port: 8080
spring:
  servlet:
    multipart:
      enabled: true
      #最大支持文件大小
      max-file-size: 100MB
      #最大支持请求大小
      max-request-size: 100MB
```

接下来创建文件上传的处理接口 FileController 类，代码如下：

```java
@Controller
@RequestMapping("/file/")
public class FileController {
    /*单文件上传*/
    @RequestMapping("upload")
    @ResponseBody
    public String upload(@RequestParam("file") MultipartFile file) {
        //获取原始名字
        String fileName = file.getOriginalFilename();
        //获取后缀名
        // String suffixName = fileName.substring(fileName.lastIndexOf("."));
        //文件保存路径
        String filePath = "d:/file/";
        //文件重命名,防止重复
        fileName = filePath + UUID.randomUUID() + fileName;
        //文件对象
```

```
            File dest = new File(fileName);
            //判断路径是否存在,如果不存在则创建
            if(!dest.getParentFile().exists()) {
                dest.getParentFile().mkdirs();
            }
            try {
                //保存到服务器中
                file.transferTo(dest);
                return "上传成功";
            } catch (Exception e) {
                e.printStackTrace();
            }
            return "上传失败";
        }
    }
```

Spring Boot 项目启动后,在浏览器中输入:http://localhost:8080/upload.html 进行文件上传,如图 12-1 所示。

单击"选择文件"按钮上传文件,文件上传成功后,会在页面上提示上传成功,并且在计算机的 D:/file/中存在一个刚刚上传的文件,如图 12-2 所示。

图 12-1 文件上传

图 12-2

12.1.2 多文件上传

多个文件的上传和单个文件的上传方法基本一致,只是在控制器里多一个遍历的步骤。首先在 upload.html 文件中添加多文件上传代码,代码如下:

```
<p>多文件上传</p>
<form method="POST" enctype="multipart/form-data" action="/file/uploads">
    <!-- 添加 multiple 属性,可以按住 ctrl 键多选文件 -->
    <p>文件:<input type="file" multiple name="file" /></p>
    <p><input type="submit" value="上传" /></p>
</form>
```

然后在后台控制器中添加多文件上传的逻辑代码,代码如下:

```java
/*多文件上传*/
@PostMapping("/uploads")
@ResponseBody
public String handleFileUpload(HttpServletRequest request) {
    List<MultipartFile> files = ((MultipartHttpServletRequest) request)
            .getFiles("file");
    MultipartFile file = null;
    for (int i = 0; i < files.size(); ++i) {
        file = files.get(i);
        if (!file.isEmpty()) {
            try {
                //获取原始名字
                String fileName = file.getOriginalFilename();
                //文件保存路径
                String filePath = "d:/file/";
                //文件重命名,防止重复
                fileName = filePath + UUID.randomUUID() + fileName;
                //文件对象
                File dest = new File(fileName);
                //保存到服务器
                file.transferTo(dest);
            } catch (Exception e) {
                e.printStackTrace();
                return "上传失败";
            }
        }else {
            return "上传失败";
        }
    }
    return "上传成功";
}
```

这样就可以实现多个文件同时上传的效果。

12.1.3 文件下载

文件下载,这个时候不再需要创建 html 文件了,可以直接输入网址进行下载,接着看一看下载核心代码,也是非常简单的,代码如下:

```java
@RequestMapping("download")
public void download(HttpServletResponse response) throws Exception {
    //文件地址,真实环境是存放在数据库中
    String filename = "a.txt";
    String filePath = "D:/file";
    File file = new File(filePath + "/" + filename);
```

```java
        //创建输入对象
        FileInputStream fis = new FileInputStream(file);
        //设置相关格式
        response.setContentType("application/force-download");
        //设置下载后的文件名及header
        response.setHeader("Content-Disposition", "attachment;fileName=" + filename);
        //创建输出对象
        OutputStream os = response.getOutputStream();
        //常规操作
        byte[] buf = new byte[1024];
        int len = 0;
        while((len = fis.read(buf)) != -1) {
            os.write(buf, 0, len);
        }
        fis.close();
    }
```

启动项目，当我们在浏览器的地址栏中输入 http://localhost:8080/file/download 时会发现，文件可以下载了。

12.2 定时器

定时任务，在企业开发中尤其重要，很多业务都需要定时任务去完成。例如 10 点开售某件商品，凌晨 0 点统计注册人数，以及统计其他各种信息等。这个时候不可能人为地去开启某个开关或者人为地统计某些数据，如果这样，估计执行任务者都要崩溃了。

Spring Boot 定时器的使用一般有以下几种实现方式。

（1）Timer：这是 Java 自带的 java.util.Timer 类，这个类允许调度一个 java.util.TimerTask 任务。使用这种方式可以让程序按照某一个频率执行，但不能在指定时间运行。一般用得较少。

（2）ScheduledExecutorService：也 jdk 自带的一个类，是基于线程池设计的定时任务类，每个调度任务都会分配到线程池中的一个线程去执行，也就是说任务是并发执行的，它们之间互不影响。

（3）Spring Task：Spring 3.0 以后自带的 Task，可以将它看成一个轻量级的 Quartz，而且使用起来比 Quartz 简单许多。

（4）Quartz：这是一个功能比较强大的的调度器，可以让程序在指定的时间执行，也可以按照某一个频率执行，配置起来稍显复杂。

简单的定时任务可以直接通过 Spring 自带的 task 来实现，复杂的定时任务则可以通过集成的 Quartz 来实现，下面分别进行介绍。

12.2.1 Task

Task 是 Spring 自带的定时器,使用方便、简单。使用方式如下。

1. 创建工程

首先创建一个 Spring Boot Web 工程项目。

2. 开启定时任务

在启动类添加@EnableScheduling 注解,开启对定时任务的支持,代码如下:

```
@SpringBootApplication
@EnableScheduling
public class TimerSetApplication {

    public static void main(String[] args) {
        SpringApplication.run(TimerSetApplication.class, args);
    }
}
```

3. 设置定时任务

定时任务主要通过 @Scheduled 注解来实现,代码如下:

```
@Component
public class ScheduledTask {
    private Logger log = LoggerFactory.getLogger(ScheduledTask.class);
    @Scheduled(cron = "0 0/1 * * * ?")
    public void testOne() {
        log.info("每分钟执行一次");
    }

    @Scheduled(fixedRate = 30000)
    public void testTwo() {
        log.info("每30s执行一次");
    }
    @Scheduled(cron = "0 0 1 * * ?") //表示每天凌晨一点执行
    public void initTask() {
        //执行任务
        log.info("执行任务" + new Date());
    }
}
```

配置完成后,启动 Spring Boot 项目,观察控制台日志打印信息,如图 12-3 所示。

4. cronExpression 表达式

常用的 cronExpression 表达式如表 12-1 所示。

```
INFO 17844 --- [   scheduling-1] com.beixi.timer.ScheduledTask            : 每30s执行一次
INFO 17844 --- [   scheduling-1] com.beixi.timer.ScheduledTask            : 每30s执行一次
INFO 17844 --- [   scheduling-1] com.beixi.timer.ScheduledTask            : 每分钟执行一次
INFO 17844 --- [   scheduling-1] com.beixi.timer.ScheduledTask            : 每30s执行一次
INFO 17844 --- [   scheduling-1] com.beixi.timer.ScheduledTask            : 每30s执行一次
INFO 17844 --- [   scheduling-1] com.beixi.timer.ScheduledTask            : 每分钟执行一次
INFO 17844 --- [   scheduling-1] com.beixi.timer.ScheduledTask            : 每30s执行一次
INFO 17844 --- [   scheduling-1] com.beixi.timer.ScheduledTask            : 每30s执行一次
INFO 17844 --- [   scheduling-1] com.beixi.timer.ScheduledTask            : 每分钟执行一次
```

图 12-3　定时器日志信息

表 12-1　表达式说明

字　　段	允　许　值	允许的特殊字符
秒	0-59	, - * /
分	0-59	, - * /
小时	0-23	, - * /
日期	1-31	, - * / L W C
月份	1-12 或者 JAN-DEC	, - * /
星期	1-7 或者 SUN-SAT	, - * / L C #
年(可选)	留空，1970-2099	, - * /

如上面的表达式所示：

(1)"*"字符被用来指定所有的值。如："*"在分钟的字段域里表示"每分钟"。

(2)"-"字符被用来指定一个范围。如："10-12"在小时域里表示"10 点、11 点、12 点"。

(3)","字符被用来指定另外的值。如："MON,WED,FRI"在星期域里表示"星期一、星期三、星期五"。

(4)"?"字符只在日期域和星期域中使用。它被用来指定"非明确的值"。当需要通过在这两个域中的一个来指定某日期或星期几的时候，它是有用的。看一看下面所举的例子就会明白。

(5)"L"字符指定在月或者星期中的某天(最后一天)。即"Last"的缩写。但是在星期和月中"L"表示不同的意思，如：在月子段中"L"指月份的最后一天，例如 1 月 31 日和 2 月 28 日，如果在星期字段中则简单地表示为"7"或者"SAT"。如果在星期字段中在某个 value 值的后面，则表示"某月的最后一个星期 value"，如"6L"表示某月的最后一个星期五。

(6)"W"字符只能用在月份字段中，该字段指定了离指定日期最近的那个星期日。

(7)"#"字符只能用在星期字段，该字段指定了第几个星期 value 在某月中。

每一个元素都可以显式地规定一个值(如 6)，一个区间(如 9~12)，一个列表(如 9,11,13)或一个通配符(如 *)。"月份中的日期"和"星期中的日期"这两个元素是互斥的，因此应该通过设置一个问号(?)来表明不想设置的那个字段。

注意：Cron 表达式对特殊字符的大小写不敏感，对代表星期的缩写英文大小写也不敏感。

Cron 表示式实例的格式:

| [秒] | [分] | [时] | [日] | [月] | [周] | [年] |

一些常用的 Cron 表达式示例如表 12-2 所示。

表 12-2　Cron 表达式示例

表　达　式	意　　义
"0 0 12 * * ?"	每天中午 12 点触发
"0 15 10 ? * *"	每天上午 10:15 触发
"0 15 10 * * ?"	每天上午 10:15 触发
"0 15 10 * * ? *"	每天上午 10:15 触发
"0 15 10 * * ? 2005"	2005 年的每天上午 10:15 触发
"0 * 14 * * ?"	在每天下午 2 点到下午 2:59 期间的每 1 分钟触发
"0 0/5 14 * * ?"	在每天下午 2 点到下午 2:55 期间的每 5 分钟触发
"0 0/5 14,18 * * ?"	在每天下午 2 点到 2:55 期间和下午 6 点到 6:55 期间的每 5 分钟触发
"0 0-5 14 * * ?"	在每天下午 2 点到下午 2:05 期间的每 1 分钟触发
"0 10,44 14 ? 3 WED"	每年三月的星期三的下午 2:10 和 2:44 触发
"0 15 10 ? * MON-FRI"	周一至周五的上午 10:15 触发
"0 15 10 15 * ?"	每月 15 日上午 10:15 触发
"0 15 10 L * ?"	每月最后一日的上午 10:15 触发
"0 15 10 ? * 6L"	每月的最后一个星期五上午 10:15 触发
"0 15 10 ? * 6L 2002-2005"	2002 年至 2005 年的每月的最后一个星期五上午 10:15 触发
"0 15 10 ? * 6#3"	每月的第 3 个星期五上午 10:15 触发

12.2.2　Quartz

1. Quartz 简介

Spirng Boot 可以使用 Quartz 实现定时器的功能,是一个完全由 Java 编写的开源任务调度框架,通过触发器设置作业定时运行规则、控制作业的运行时间。Quartz 定时器作用有很多,例如定时发送信息、定时生成报表等。

Quartz 框架主要核心组件包括调度器、触发器、作业。调度器作为作业的总指挥,触发器作为作业的操作者,作业为应用的功能模块,组件之间的关系如图 12-4 所示。

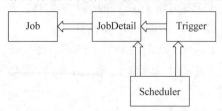

图 12-4　Quartz 各个组件关系

(1) Job 表示一个任务(工作),要执行的具体内容。

(2) JobDetail 表示一个具体的可执行的调度程序,Job 是这个可执行调度程序所要执行的内容,另外 JobDetail 还包含了这个任务调度的方案和策略。

告诉调度容器,将来执行哪个类(job)的哪个方法。

(3) Trigger 是一个类,代表一个调度参数的配置,描述触发 Job 执行的时间及触发规则。一个 Job 可以对应多个 Trigger,但一个 Trigger 只能对应一个 Job。

(4) Scheduler 代表一个调度容器,一个调度容器中可以注册多个 JobDetail 和 Trigger。Scheduler 可以将 Trigger 绑定到某一个 JobDetail 中,这样当 Trigger 触发时,对应的 Job 就会被执行。

当 JobDetail 和 Trigger 在 Scheduler 容器上注册后,形成了装配好的作业(JobDetail 和 Trigger 所组成的一对儿),这样就可以伴随容器启动而调度执行了。

2. 整合 Spring Boot

添加 Quartz 的依赖,代码如下:

```xml
<!-- 定时器依赖 -->
<dependency>
    <groupId>org.quartz-scheduler</groupId>
    <artifactId>quartz</artifactId>
    <version>2.2.1</version>
</dependency>
<!-- 该依赖必加,里面有 Sping 对 Schedule 的支持 -->
<dependency>
    <groupId>org.springframework</groupId>
    <artifactId>spring-context-support</artifactId>
</dependency>
<!-- 必须添加,要不然会出错,导致项目无法启动 -->
<dependency>
    <groupId>org.springframework</groupId>
    <artifactId>spring-tx</artifactId>
</dependency>
```

任务执行类(任务执行,并且通过控制器的接口实现时间间隔的动态修改任务类),代码如下:

```java
@Configuration
@Component
@EnableScheduling
public class JobTask {
    public void start() throws InterruptedException {
        SimpleDateFormat format = new SimpleDateFormat("yyyy-MM-dd HH:mm:ss");
        System.err.println("定时任务开始执行。" + format.format(new Date()));
    }
}
```

Quartz 的详细配置类,代码如下:

```java
@Configuration
public class QuartzConfigration {
    @Bean(name = "jobDetail")
    public MethodInvokingJobDetailFactoryBean detailFactoryBean(JobTask task) {
        //ScheduleTask 为需要执行的任务
        MethodInvokingJobDetailFactoryBean jobDetail = new MethodInvokingJobDetailFactoryBean();
        /*
         * 是否并发执行
         * 例如每 3s 执行一次任务,但是当前任务还没有执行完,就已经过了 3s 了。
         * 如果此处为 true,则下一个任务会并发执行,如果此处为 false,则下一个任务会等待
         上一个任务执行完后,再开始执行
         */
        jobDetail.setConcurrent(true);
        jobDetail.setName("scheduler");         //设置任务的名字
        jobDetail.setGroup("scheduler_group");//设置任务的分组,这些属性都可以存储在数据库
                                              //中,在多任务的时候使用


        /*
         * 这两行代码表示执行 task 对象中的 scheduleTest 方法。定时执行的逻辑都在
         scheduleTest 中。
         */
        jobDetail.setTargetObject(task);
        jobDetail.setTargetMethod("start");
        return jobDetail;
    }

    @Bean(name = "jobTrigger")
      public CronTriggerFactoryBean cronJobTrigger ( MethodInvokingJobDetailFactoryBean
        jobDetail) {
        CronTriggerFactoryBean tigger = new CronTriggerFactoryBean();
        tigger.setJobDetail(jobDetail.getObject());
        tigger.setCronExpression("*/5 * * * * ?");         //每 5s 执行一次
          tigger.setName("myTigger");                      // trigger 的 name
        return tigger;
    }
    @Bean(name = "scheduler")
    public SchedulerFactoryBean schedulerFactory(Trigger cronJobTrigger) {
        SchedulerFactoryBean bean = new SchedulerFactoryBean();
        //设置是否任意一个已定义的 Job 会覆盖现在的 Job。默认为 false,即已定义的 Job 不
        //会覆盖现有的 Job
        bean.setOverwriteExistingJobs(true);
                //延时启动,应用启动 5s 后,定时器才开始启动
        bean.setStartupDelay(5);
                //注册定时触发器
        bean.setTriggers(cronJobTrigger);
```

```
            return bean;
    }
    //多任务时的Scheduler,动态设置Trigger。一个SchedulerFactoryBean可能会有多个Trigger
    @Bean(name = "multitaskScheduler")
    public SchedulerFactoryBean schedulerFactoryBean(){
        SchedulerFactoryBean schedulerFactoryBean = new SchedulerFactoryBean();
        return schedulerFactoryBean;
    }
}
```

到此为止,定时任务就配置成功了。接下来启动Spring Boot项目,查看控制台日志信息,如图12-5所示。

```
2020-08-18 22:29:12.591  INFO 6540 ---
定时任务开始执行。2020-08-18 22:29:12
定时任务开始执行。2020-08-18 22:29:15
定时任务开始执行。2020-08-18 22:29:20
定时任务开始执行。2020-08-18 22:29:25
定时任务开始执行。2020-08-18 22:29:30
定时任务开始执行。2020-08-18 22:29:35
定时任务开始执行。2020-08-18 22:29:40
```

图12-5　日志信息

12.3　Spring Boot发送Email

在开发中,经常会碰到Email邮件发送的场景,如注册、找回密码、发送验证码、向客户发送邮件、通过邮件发送系统情况、通过邮件发送报表信息等,实际应用场景很多。

首先介绍以下与发送和接收邮件相关的一些协议。

(1) 发送邮件:SMPT、MIME,是一种基于"推"的协议,通过SMPT协议将邮件发送至邮件服务器,MIME协议是对SMPT协议的一种补充,如发送图片附件等。

(2) 接收邮件:POP、IMAP,是一种基于"拉"的协议,收件人通过POP协议从邮件服务器拉取邮件。

12.3.1　发送邮件需要的配置

因为各大邮件运营商都有其对应的安全系统,不是项目中想用就可以用的,我们必须获得其对应的客户端授权码才行,获得授权码后,在项目中配置SMTP服务协议及主机配置账户,这样就可以在项目中使用各大邮件运营商进行发送邮件了。

由于国内使用QQ邮箱的用户较多,所以这里选择QQ邮箱为例。具体实现需要登录QQ邮箱,单击"设置"按钮,然后选择账户选项,向下拉并选择开启POP3/SMTP服务,如图12-6所示。

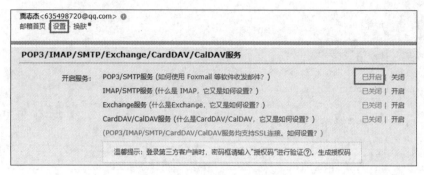

图 12-6　开启 POP3/SMTP 服务

单击"开启"按钮后会进入验证过程，根据引导步骤发送短信，验证成功后即可得到自己 QQ 邮箱的客户端授权码了，如图 12-7 所示。

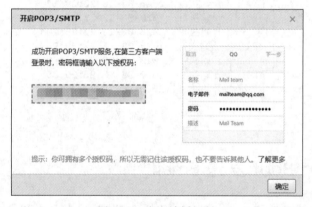

图 12-7　客户端授权码

我们在获得授权码后，就可以在 Spring Boot 工程中配置文件 aplication.yml 或者 properties 文件中进行配置了。

12.3.2　使用 Spring Boot 发送邮件

相信使用过 Spring 的众多开发者都知道 Spring 提供了非常好用的 JavaMailSender 接口实现邮件发送。在 Spring Boot 的 Starter 模块中也为此提供了自动化配置。下面通过实例看一看如何在 Spring Boot 中使用 JavaMailSender 发送邮件。

1．环境搭建

首先创建 Spring Boot 项目，需要引入如下 Email 依赖：

```xml
<dependency>
    <groupId>org.springframework.boot</groupId>
    <artifactId>spring-boot-starter-mail</artifactId>
</dependency>
```

然后在配置文件 application.yml 中加入相关配置信息：

```yaml
## QQ邮箱配置
spring:
  mail:
    host: smtp.qq.com                    # 发送邮件服务器
    username: 635498720@qq.com           # 发送邮件的邮箱地址
    password: (12.3.1 中获取的授权码)     # 客户端授权码,不是邮箱密码
    smtp.port: 465                       # 端口号 465 或 587
    from: 635498720@qq.com               # 发送邮件的地址和上面 username 的邮箱地址一致
## 可以任意
    properties.mail.smtp.starttls.enable: true
    properties.mail.smtp.starttls.required: true
    properties.mail.smtp.ssl.enable: true
    default-encoding: utf-8
```

2．发送普通邮件

普通邮件是指最为普通的纯文本邮件。创建 MailService 类用来封装邮件的发送，代码如下：

```java
@Service
public class MailService {
    private final Logger logger = LoggerFactory.getLogger(this.getClass());

    /* Spring Boot 提供了一个发送邮件的简单抽象,使用的是下面这个接口,这里直接注入即可使用 */
    @Autowired
    private JavaMailSender mailSender;
    /* 获取配置文件中自己的 QQ 邮箱 */
    @Value("${spring.mail.from}")
    private String from;

    /**
     * 简单文本邮件
     * @param to 收件人
     * @param subject 主题
     * @param content 内容
     */
    public void sendSimpleMail(String to, String subject, String content) {
        //创建 SimpleMailMessage 对象
        SimpleMailMessage message = new SimpleMailMessage();
        //邮件发送
        message.setFrom(from);
        //邮件接收人
        message.setTo(to);
        //邮件主题
```

```
        message.setSubject(subject);
        //邮件内容
        message.setText(content);
        //通过JavaMailSender类把邮件发送出去
        mailSender.send(message);
    }
}
```

代码解释：

（1）JavaMailSender 是 Spring Boot 提供的邮件发送接口，直接注入即可使用。

（2）简单邮件通过 SimpleMailMessage 对象封装打包数据，最后通过 JavaMailSender 类将数据发送出去。

配置完成后，在 Spring Boot 提供的主测试类中进行测试，代码如下：

```
@RunWith(SpringRunner.class)
@SpringBootTest
public class EmailApplicationTests {

    /* 注入发送邮件的接口 */
    @Autowired
    private MailService mailService;
    /** 测试发送文本邮件 */
    @Test
    public void sendmail() {
        mailService.sendSimpleMail("752587470@qq.com","主题：普通邮件","内容：第一封纯文本邮件");
    }
}
```

注意：

（1）添加完毕后，在测试类中添加@RunWith(SpringRunner.class)并导入相应的包，需要注意的是，@Test 的包是导入 org.junit.Test 而不是 org.junit.jupiter.api.Test。

（2）将类和方法设置为 public 才可以运行。

执行该方法，即可看到邮件发送成功，如图 12-8 所示。

图 12-8 普通邮件

3. 发送 HTML 邮件

很多时候，需要邮件带有美观的样式。这时候，可以使用 HTML 的样式。我们需要使用 javaMailSender 的 createMimeMessage 方法，构建一个 MimeMessage，然后使用 MimeMessage 实例创建出 MimeMessageHelper。在 MailService 中添加如下方法：

```java
public void sendHtmlMail(String to, String subject, String content) {
    //获取 MimeMessage 对象
    MimeMessage message = mailSender.createMimeMessage();
        MimeMessageHelper messageHelper;
        try {
    messageHelper = new MimeMessageHelper(message, true);
    //邮件发送人
    messageHelper.setFrom(from);
    //邮件接收人
    messageHelper.setTo(to);
    //邮件主题
    message.setSubject(subject);
    //邮件内容,HTML 格式
    messageHelper.setText(content, true);
    //发送
    mailSender.send(message);
    //日志信息
    logger.info("邮件已经发送。");
    } catch (Exception e) {
        logger.error("发送邮件时发生异常!", e);
    }
}
```

在测试类中添加如下方法进行测试：

```java
@Test
public void sendmailHtml(){
    mailService.sendHtmlMail("752587470@qq.com","主题: html 邮件","<h1>内容：第一封 html 邮件</h1>");
}
```

执行该方法，即可看到邮件发送成功，如图 12-9 所示。

4. 带附件的邮件

很多时候，在发送邮件的时候，需要附带一些附件一起发送，那么在 JavaMailSender 中也有附带附件的方法，在 MailService 中添加如下方法：

图 12-9 HTML 邮件

```java
public void sendAttachmentMail(String to, String subject, String content, String filePath) {
    logger.info("发送带附件邮件开始: {},{},{},{}", to, subject, content, filePath);
    MimeMessage message = mailSender.createMimeMessage();
    MimeMessageHelper helper;
```

```
    try {
        helper = new MimeMessageHelper(message, true);
        //true 代表支持多组件,如附件、图片等
        helper.setFrom(from);
        helper.setTo(to);
        helper.setSubject(subject);
        helper.setText(content, true);
        FileSystemResource file = new FileSystemResource(new File(filePath));
        String fileName = file.getFilename();
//添加附件,可多次调用该方法添加多个附件
        helper.addAttachment(fileName, file);
        mailSender.send(message);
        logger.info("发送带附件邮件成功");
    } catch (MessagingException e) {
        logger.error("发送带附件邮件失败", e);
    }
}
```

在测试类中添加如下方法进行测试:

```
@Test
public void sendAttachmentMail() {
    String content = "<html><body><h3><font color=\"red\">" + "大家好,这是 Spring Boot 发送的 HTML 邮件,有附件哦" + "</font></h3></body></html>";
    String filePath = "E:\\C 语言书籍\\第 1 章 C 语言概述.docx";
    mailService.sendAttachmentMail("752587470@qq.com", "发送邮件测试", content, filePath);
}
```

执行该方法,即可看到邮件发送成功,如图 12-10 所示。

图 12-10 带附件的邮件

5. 发送带图片的邮件

带图片即在正文中使用<img>标签,并设置我们需要发送的图片,在 HTML 基础上添加一些参数即可,在 MailService 中添加如下方法:

```java
public void sendInlineResourceMail(String to, String subject, String content, String
        rscPath, String rscId) {

    logger.info("发送带图片邮件开始: {},{},{},{},{}", to, subject, content, rscPath,
rscId);
    MimeMessage message = mailSender.createMimeMessage();

    MimeMessageHelper helper;
    try {
        helper = new MimeMessageHelper(message, true);
        helper.setFrom(from);
        helper.setTo(to);
        helper.setSubject(subject);
        helper.setText(content, true);
        FileSystemResource res = new FileSystemResource(new File(rscPath));
        helper.addInline(rscId, res);        //重复使用可添加多张图片
        mailSender.send(message);
        logger.info("发送带图片邮件成功");
    } catch (MessagingException e) {
        logger.error("发送带图片邮件失败", e);
    }
}
```

在测试类中添加如下方法进行测试:

```java
@Test
public void sendInlineResourceMail() {
    String rscPath = "D:\\cat.jpg";
    String rscId = "001";
    /*使用 cid 标注出静态资源*/
    String content = "<html><body><h3><font color=\"red\">" + "大家好,这是 Spring
            Boot 发送的 HTML 邮件,有图片哦" + "</font></h3>"
            + "<img src=\'cid:" + rscId + "\'></body></html>";
    mailService.sendInlineResourceMail("752587470@qq.com", "发送邮件测试", content,
            rscPath, rscId);
}
```

执行该方法,即可看到邮件发送成功,如图 12-11 所示。

6. 模板邮件

通常我们使用邮件发送所需服务信息的时候,都会有一些固定的场景,例如重置密码、

图 12-11 发送带图片的邮件

注册确认等,给每个用户发送的内容可能只有小部分是变化的。所以,很多时候我们会使用模板引擎将各类邮件设置成模板,这样只需在发送时替换变化部分的参数。

在 Spring Boot 中使用模板引擎实现模板化的邮件发送也是非常容易的,下面我们以 Thymeleaf 为例实现模板化的邮件发送。

首先添加 Thymeleaf 的依赖,代码如下:

```xml
<dependency>
    <groupId>org.springframework.boot</groupId>
    <artifactId>spring-boot-starter-thymeleaf</artifactId>
</dependency>
```

在 template 文件夹下创建 emailTemplate.html,代码如下:

```html
<!DOCTYPE html>
<html lang="en" xmlns:th="http://www.thymeleaf.org">
<head>
<meta charset="UTF-8">
<title>邮件模板</title>
</head>
<body>
    您好,感谢您的注册,这是一封验证邮件,请单击下面的链接完成注册,感谢您的支持!<br>
    <a href="#" th:href="@{http://www.bestbpf.com/register/{id}(id=${id})}">激活账户</a>
</body>
</html>
```

在模板页面中,id 是动态变化的,需要传参设置,其实就是传参后,将页面解析为 HTML 字符串,作为邮件发送的主体内容。在测试类中添加如下代码进行测试:

```
/* TemplateEngine 来对模板进行渲染 */
@Autowired
private TemplateEngine templateEngine;
@Test
public void testTemplateMail() {
    //向 Thymeleaf 模板传值,并解析成字符串
    Context context = new Context();
    context.setVariable("id", "001");
    String emailContent = templateEngine.process("emailTemplate", context);

    mailService.sendHtmlMail("752587470@qq.com", "这是一个模板文件", emailContent);
}
```

执行该方法,即可看到邮件发送成功,如图 12-12 所示。

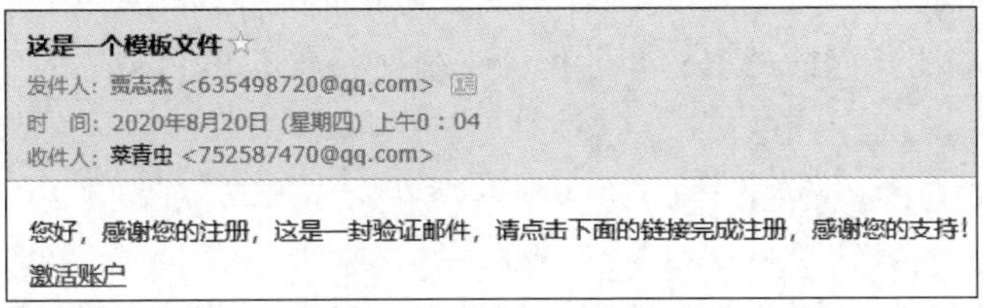

图 12-12　模板邮件

第 13 章 Spring Boot 热部署和 Postman 工具

在 Spring Boot 中提供了 spring-boot-devtools 开发工具,实现了 Spring Boot 的热部署,热部署就是在项目正在运行的时候修改代码,却不需要重新启动项目,大大提高了开发效率。

前后端分离太有用了,对于后端来说,不用去考虑前端的布局,只需要考虑后端数据的正确性,那么在后端测试返回的 JSON 数据的正确性就离不开 Postman 工具。

13.1 devtools 热部署

在实际开发过程中,如果每次修改代码就得将项目重启,重新部署,则对于一些大型应用来说,重启需要花费大量的时间成本。对于一个后端开发者来说,重启过程确实很难受。在 Spring Boot 中使用 devtools 工具包实现代码热部署是一件很简单的事情,代码的修改可以自动部署并重新热启动项目。

13.1.1 热部署原理

在项目开发过程中,常常会改动页面数据或者修改数据结构,为了显示改动效果,往往需要重启应用并查看改变后的效果,其实就是重新编译生成了新的 Class 文件,在这个文件里记录着和代码等对应的各种信息,然后 Class 文件将被虚拟机的 ClassLoader 加载。

而热部署正是利用了这个特点,如果它监听到有 Class 文件改动了,就会创建一个新的 ClaassLoader 进行加载该文件,经过一系列的过程,最终将结果呈现在我们眼前。

深层原理是使用两个 ClassLoader,一个 ClassLoader 加载那些不会改变的类(第三方 Jar 包),另一个 ClassLoader 加载会更改的类,称为 restart ClassLoader,这样在有代码更改的时候,原来的 restart ClassLoader 被丢弃,重新创建一个 restart ClassLoader,由于需要加载的类相对较少,所以缩短了重启时间。

13.1.2 devtools 应用

1. 添加依赖

首先在 pom.xml 文件中添加 devtools 依赖,代码如下:

```xml
<dependency>
    <groupId>org.springframework.boot</groupId>
    <artifactId>spring-boot-devtools</artifactId>
    <optional>true</optional>
</dependency>
```

说明事项：

（1）devtools可以实现页面热部署，即页面修改后会立即生效，这个可以直接在application.properties文件中配置spring.thymeleaf.cache=false来实现。

实现类文件热部署，实现对属性文件的热部署，即devtools会监听classpath下的文件变动，并且会立即重启应用，因为其采用的是虚拟机机制，该项重启是很快的。

（2）配置了true后在修改Java文件后也就支持了热启动，不过这种方式属于项目重启（速度比较快的项目重启），会清空session中的值，也就是如果有用户登录，项目重启后用户需要重新登录。

默认情况下，/META-INF/maven、/META-INF/resources、/resources、/static、/templates、/public这些文件夹下的文件修改后不会使应用重启，但是会重新加载，devtools内嵌了一个LiveReload server，当资源发生改变时，浏览器刷新。

2. application.yml中配置devtools

以下配置用于自定义配置热部署，也可不设置，代码如下：

```yaml
spring:
  devtools:
    restart:
      enabled: true                          #设置开启热部署
      additional-paths: src/main/java        #重启目录
      exclude: WEB-INF/**                    #指定目录不更新
  freemarker:
    cache: false                             #页面不加载缓存,修改即时生效
```

3. IDEA中配置

当我们修改了类文件后，IDEA不会自动编译，得修改IDEA设置。

（1）File-Settings-Compiler-Build Project automatically，如图13-1所示。

（2）Ctrl+Shift+Alt+/，选择Registry，勾上compiler.automake.allow.when.app running，如图13-2所示。

通过以上步骤，就完成了Spring Boot项目的热部署功能。

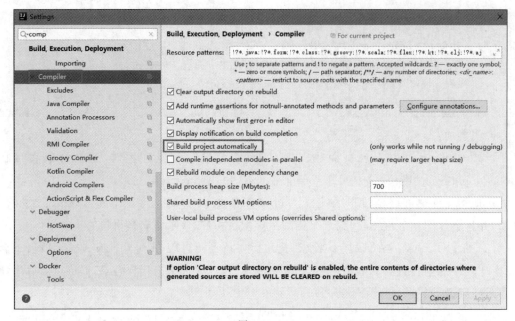

图 13-1

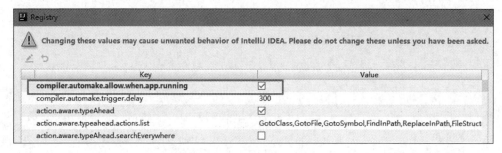

图 13-2

4. 测试

对热部署进行测试,代码如下:

```
@RestController
public class HelloDemo {
    @RequestMapping("/index")
    public String index() {
        return "index!";
    }
}
```

启动项目,通过浏览器输入地址:http://localhost:8080/index,结果如图 13-3 所示。新添加请求,在不重新启动项目的情况下测试热部署是否配置成功,代码如下:

```
@RestController
public class HelloDemo {
    @RequestMapping("/index")
    public String index() {
        return "index!";
    }

    @RequestMapping("/main")
    public String home(){
        return "home!";
    }
}
```

测试新加请求是否成功,浏览器输入 http://localhost:8080main 后结果如图 13-4 所示。

图 13-3　　　　　　　　　　　　　　图 13-4

至此说明热部署配置生效了。

13.2　Postman 工具

13.2.1　Postman 介绍

Postman 是一款功能强大的网页调试与发送网页 HTTP 请求的工具。Postman 能够发送任何类型的 HTTP 请求(GET、HEAD、POST、PUT…),附带任何数量的参数和 HTTP headers。支持不同的认证机制(basic、digest、OAuth),接收到的响应语法高亮(HTML、JSON 或 XML)。Postman 既可以以 Chrome 浏览器插件的形式存在,也可以独立的应用程序存在。本节以本地安装为例来进行讲解。

使用 Spring Boot 开发 Web 项目一般有两种类型,一种是传统的前后端在同一个项目(jsp、freemarker 等),另一种是前后端分离的项目(API 形式,包括 App)。本节以基于 API 形式的接口调用来介绍 Postman 这款工具的使用。

13.2.2　Postman 下载安装

官方网站:https://www.getpostman.com/apps。

Postman 适用于不同的操作系统,Postman Mac、Windows x32、Windows x64、Linux 系统,还支持 Postman 浏览器扩展程序、Postman Chrome 应用程序等。根据自己的情况下载并安装对应环境的版本。

下载后，按提示进行默认安装，这里就省略了。安装完成后，首页界面及基础功能如图 13-5 所示。

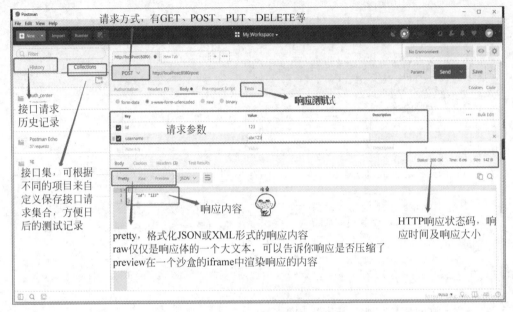

图 13-5　Postman 首页及基础功能

13.2.3　Spring Boot 基于 Postman 的 RESTful 接口调用

首先创建一个 Web 项目，提供了一个登录的 API 请求。这里简单模拟，请求参数只有用户名和密码，后台直接返回的结果为 JSON 字符串，代码如下：

```
@RestController
public class PostmanController {
    @PostMapping("login")
    public JsonResult login(String username, String password){
        JsonResult result = new JsonResult();
        result.setData(username);
        result.setCode("0");
        result.setMsg("操作成功!");
        return result;
    }
}
```

代码解释：代码中使用的 JsonResult 类是 11.3.3 节中统一封装的 JSON 结构。

创建模拟请求之后便可以填写请求相关信息和进行请求测试了，如图 13-6 所示。

由于上面的实例需要采用 POST 请求，因此在请求链接前面选择 POST，当然 Postman 还支持更多类型的请求。

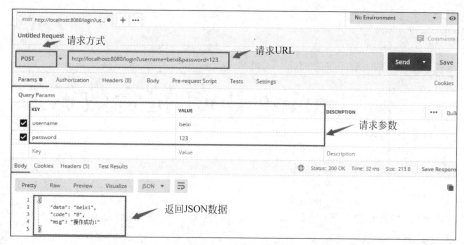

图 13-6　POST 请求

当每次请求完成之后,在左边的 History 中便会记录之前请求的历史记录。

如果请求比较多,则可通过左边的 Collection 进行分类,在不同的类别中创建具体的请求。

关于返回结果,直接以 JSON 结构数据高亮形式展示。同时,也展示了 HTTP 请求常见的 Cookies、Headers 等信息。如果返回的结果不是 JSON 结构数据,而是 XML、HTML、Text 等,Postman 也会自动解析出来。

在上面的请求中,虽然我们选择了 POST 方式提交,但是我们配置参数是通过 Params 进行配置的,此时参数依旧会被拼接到 URL 上。如果我们的请求需要通过 Header、Body 传输参数,Postman 同样支持,具体功能位于 Params 选择的后面。下面以真正 POST 方式传递参数,如图 13-7 所示。

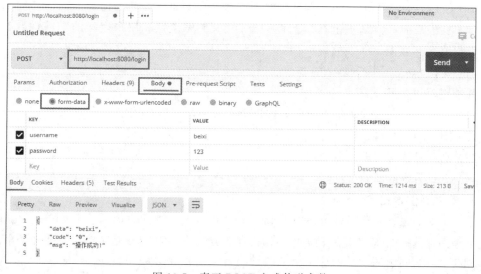

图 13-7　真正 POST 方式传递参数

第 14 章 Spring Boot 整合数据库

数据库的类型有很多，例如 MySQL、Oracle 等是关系型数据库，Redis、MongoDB 等是非关系型数据库。本章主要介绍关系型数据库：MySQL，以及非关系型数据库：Redis、MongoDB。

14.1 非关系型数据库和关系型数据库的区别

1. 关系型数据库

常见的关系型数据库有：Oracle、DB2、Microsoft SQL Server、Microsoft Access、MySQL 等。关系型数据库最典型的数据结构是表，由二维表及其之间的联系所组成的一个数据组织。

其优点有以下几点。

（1）容易理解：二维表结构非常贴近逻辑世界，关系模型相对网状、层次等其他模型来说更容易理解。

（2）使用方便：通用的 SQL 语言使得操作关系型数据库非常方便。

（3）易于维护：丰富的完整性(实体完整性、参照完整性和用户定义的完整性)大大降低了数据冗余和数据不一致的概率。

（4）支持 SQL，可用于复杂的查询。

其缺点有以下几点：

（1）为了维护一致性所付出的巨大代价就是其读写性能比较差。

（2）固定的表结构。

（3）高并发读写需求。

（4）海量数据的高效率读写。

2. 非关系型数据库

常见的非关系型数据库有 NoSQL、Cloudant、MongoDB、Redis、HBase。非关系型数据库严格上不是一种数据库，应该是一种数据结构化存储方法的集合，可以是文档或者键值对等。

其优点有以下几点。

（1）格式灵活：存储数据的格式可以是 key、value 形式、文档形式、图片形式等，使用灵活，应用场景广泛，而关系型数据库则只支持基础类型。

（2）速度快：NoSQL可以使用硬盘或者随机存储器作为载体，而关系型数据库只能使用硬盘。

（3）高扩展性。

（4）成本低：NoSQL数据库部署简单，所用软件基本都是开源软件。

其缺点有以下几点。

（1）数据和数据之间没有关系，所以不能一目了然。

（2）没有关系，没有强大的事务保证数据的完整和安全。

非关系型数据库的分类：

（1）键值（Key-Value）存储数据库：这一类数据库主要使用哈希表，在这个表中有一个特定的键和一个指针指向特定的数据。Key-Value模型对于IT系统来说优势在于简单、易部署。但是如果DBA只对部分值进行查询或更新的时候，Key-Value就显得效率低下了。这类数据库有TokyoCabinet、Redis等。

（2）列存储数据库：这部分数据库通常用来应对分布式存储的海量数据。键仍然存在，但是它们的特点是指向了多个列。这些列是由列家族来安排的。这类数据库有Cassandra、HBase、Riak等。

（3）文档型数据库：文档型数据库的灵感来自于Lotus Notes办公软件，它同第一种键值存储相类似。该类型的数据模型是版本化的文档，半结构化的文档以特定的格式存储，例如JSON。文档型数据库可以看作键值数据库的升级版，允许之间嵌套键值。而且文档型数据库比键值数据库的查询效率更高。这类数据库有CouchDB、MongoDB等。

（4）图形（Graph）数据库：图形结构的数据库与其他行列及刚性结构的SQL数据库不同，它使用灵活的图形模型，并且能够扩展到多个服务器上。NoSQL数据库没有标准的查询语言（SQL），因此进行数据库查询需要制订数据模型。许多NoSQL数据库都有REST式的数据接口或者查询API。这类数据库有Neo4J、InfoGrid等。

14.2 整合Redis缓冲

14.2.1 Redis简介

Redis是一个基于内存的单线程高性能Key-Value型数据库。整个数据库统统加载在内存当中进行操作，定期通过异步操作把数据库数据flash到硬盘上进行保存。因为是纯内存操作，Redis的性能非常出色，每秒可以处理超过10万次读写操作，是已知性能最高的Key-Value数据库。

Redis可以存储键和5种不同类型的值之间的映射。键的类型只能为字符串，值支持5种数据类型：string（字符串）、list（列表）、set（集合）、hash（哈希类型）、Sorted Set（有序集合）。

14.2.2 Redis 的安装

Redis 官方不支持 Windows 系统，但是微软公司自己做了一个支持 Windows 64 位系统的 Redis，所以我们可以在网络上下载 Redis 的 Windows 版本，下载地址为 https://GitHub.com/MicrosoftArchive/redis/releases。

进入网站后，不要选择 Pre-release（测试版），而是选择 Latest Release（稳定版）下载 Redis 安装包，如图 14-1 所示。

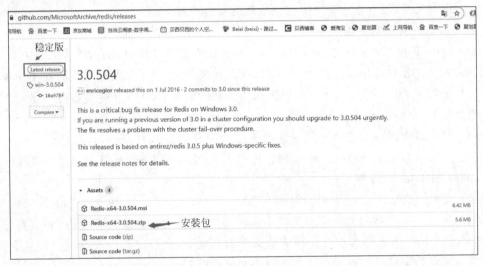

图 14-1 Windows 版本的 Redis 下载

将下载的压缩包解压到一个文件夹中，双击 redis-server.exe，Redis 服务就运行起来了，如图 14-2 所示。

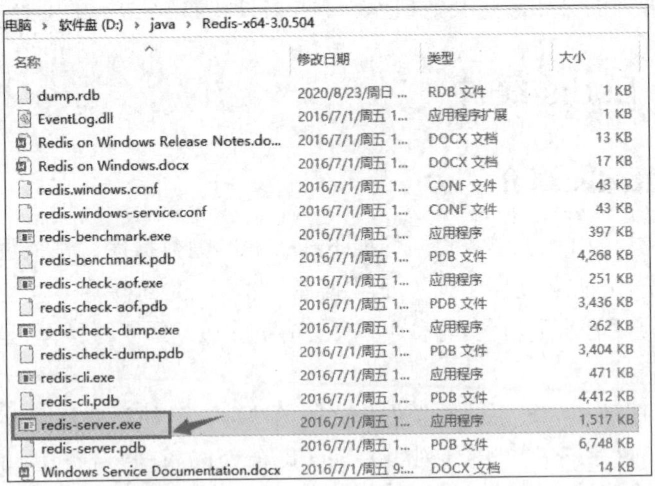

图 14-2 启动 Redis 服务器

同时我们可以看到 Redis 启动成功后的界面,如图 14-3 所示。

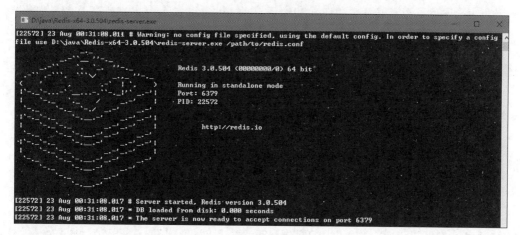

图 14-3　Redis 启动成功界面

服务器启动成功后,在安装包中双击 Redis 客户端 redis-cli.exe。启动后 Redis 客户端界面如图 14-4 所示。

图 14-4　Redis 客户端界面

14.2.3　Redis 数据库操作

1. Redis 的多数据库

Redis 默认支持 16 个数据库,这些数据库的默认命名都是从 0 开始递增的数字。

当我们连接 Redis 服务时,默认操作的是 0 号数据库,可以通过 select 命令更换数据库,如图 14-5 所示。

可以通过配置文件中的 databases 修改默认数据库个数,如图 14-6 所示。

说明:

(1) Redis 不支持自定义数据库名。

(2) Redis 不支持为每个数据库设置密码,默认数据库密码为空。
(3) Redis 的多个数据库之间不是完全隔离的,flushall 命令会清空所有数据库的数据。
(4) flushdb 命令:清空当前所在数据库的数据。

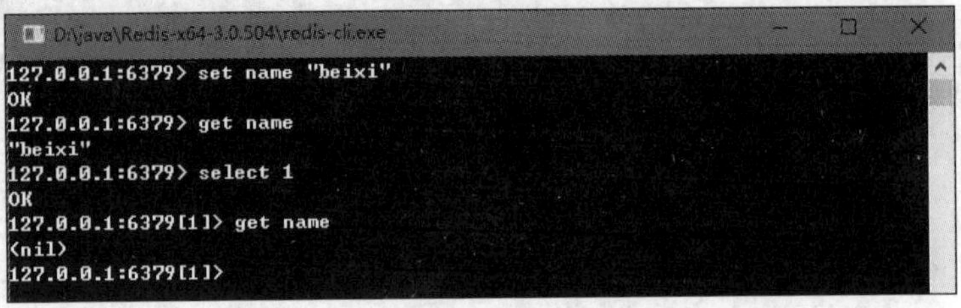

```
127.0.0.1:6379> set name "beixi"
OK
127.0.0.1:6379> get name
"beixi"
127.0.0.1:6379> select 1
OK
127.0.0.1:6379[1]> get name
<nil>
127.0.0.1:6379[1]>
```

图 14-5　切换数据库

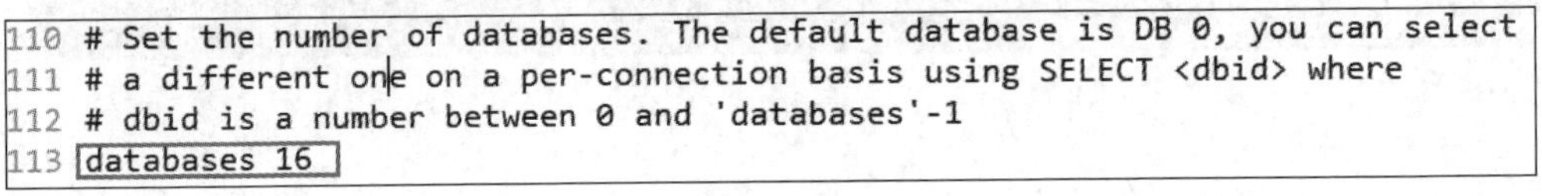

```
110 # Set the number of databases. The default database is DB 0, you can select
111 # a different one on a per-connection basis using SELECT <dbid> where
112 # dbid is a number between 0 and 'databases'-1
113 databases 16
```

图 14-6　修改默认数据库个数

2. 对字符串(string)的操作

在 Redis 数据库中对字符串的增、删、改、查操作,代码如下所示:

```
127.0.0.1:6379> set name beixi              //添加
OK
127.0.0.1:6379> get name                    //查看 values
"beixi"
127.0.0.1:6379> keys *                      //查看所有 key
1) "name"
127.0.0.1:6379> set name beixi ex 10        //设置过期时间
OK
127.0.0.1:6379> get name
"beixi"
127.0.0.1:6379> set name jzj                //更新 name 的值为 jzj
OK
127.0.0.1:6379> get name
"jzj"
127.0.0.1:6379> del name                    //删除
(integer) 0
127.0.0.1:6379> keys *
(empty list or set)
127.0.0.1:6379>
```

3. 对 List 集合的操作

在 Redis 数据库中对 List 集合的增、删、改、查操作,代码如下所示:

```
//头部插入 key 为 my_list,value 为'C++' 'java' 'web'的 list 集合
127.0.0.1:6379> lpush my_list 'C++' 'java' 'web'
(integer) 3
127.0.0.1:6379> lrange my_list 0 -1           //查询集合
1) "web"
2) "java"
3) "C++"
127.0.0.1:6379> rpush my_list 'python'        //尾部添加
(integer) 4
127.0.0.1:6379> lpush my_list 'ceshi'         //头部添加
(integer) 5
127.0.0.1:6379> lrange my_list 0 -1           //查询集合
1) "ceshi"
2) "web"
3) "java"
4) "C++"
5) "python"
127.0.0.1:6379> lset my_list 0 'ui'           //更新 index 为 0 的值
OK
127.0.0.1:6379> lrange my_list 0 -1
1) "ui"
2) "web"
3) "java"
4) "C++"
5) "python"
127.0.0.1:6379> lrem my_list 0 'ui'           //删除 index 为 0 的值
(integer) 1
127.0.0.1:6379> lrange my_list 0 -1
1) "web"
2) "java"
3) "C++"
4) "python"
127.0.0.1:6379>
```

4. 对 set 集合的操作

set 是无序的集合,其中的元素没有先后顺序。对 set 集合的操作如下所示:

```
127.0.0.1:6379> sadd my_set java c++ python   //添加元素,会自动去重
(integer) 3
127.0.0.1:6379> smembers my_set               //查询元素
1) "python"
2) "c++"
3) "java"
127.0.0.1:6379> srem my_set c++               //移除元素
```

```
(integer) 1
127.0.0.1:6379> scard my_set            //查询集合中的元素个数
(integer) 2
127.0.0.1:6379> smembers my_set
1) "python"
2) "java"
127.0.0.1:6379> sadd my_set2 c++ python
(integer) 2
127.0.0.1:6379> sunion my_set my_set2   //获取多个集合的并集
1) "python"
2) "c++"
3) "java"
127.0.0.1:6379> sinter my_set my_set2   //获取多个集合的交集
1) "python"
127.0.0.1:6379>
```

5．对 Hash 集合的操作

在 Redis 数据库中对 Hash 集合的增、删、改、查操作，代码如下：

```
//添加 key 为 my_hset,字段为 name,值为 jzj
127.0.0.1:6379> hset my_hset name jzj
(integer) 1
//在 key 为 my_hset 的哈希集中添加字段 2
127.0.0.1:6379> hset my_hset name2 jzj2
(integer) 1
//查询 my_hset 字段长度
127.0.0.1:6379> hlen my_hset
(integer) 2
//查询所有字段
127.0.0.1:6379> hkeys my_hset
1) "name"
2) "name2"
//查询所有值
127.0.0.1:6379> hvals my_hset
1) "jzj"
2) "jzj2"
//查询字段 name 的值
127.0.0.1:6379> hget my_hset name
"jzj"
//获取 key 为 my_hset 的哈希集的所有字段和值
127.0.0.1:6379> hgetall my_hset
1) "name"
2) "jzj"
3) "name2"
```

```
4) "jzj2"
//更新字段 name 的值为 new_jzj
127.0.0.1:6379> hset my_hset name new_jzj
(integer) 0
//获取 key 为 my_hset 的哈希集的所有字段和值
127.0.0.1:6379> hgetall my_hset
1) "name"
2) "new_jzj"
3) "name2"
4) "jzj2"
//删除字段 name 的值
127.0.0.1:6379> hdel my_hset name
(integer) 1
127.0.0.1:6379> hgetall my_hset
1) "name2"
2) "jzj2"
127.0.0.1:6379>
```

6. 对 zset 的操作

zset 是一种有序集合（sorted set），其中每个元素都关联一个序号 score。对 zset 的操作，代码如下：

```
//添加 baidu.com 元素，分数为 1
127.0.0.1:6379> zadd my_zset 1 'baidu.com'
(integer) 1
//添加 taobao.com 元素，分数为 2
127.0.0.1:6379> zadd my_zset 2 'taobao.com'
(integer) 1
//添加 qq.com 元素，分数为 3
127.0.0.1:6379> zadd my_zset 3 'qq.com'
(integer) 1
//按照分数由小到大查询 my_zset 集合的元素
127.0.0.1:6379> zrange my_zset 0 -1
1) "baidu.com"
2) "taobao.com"
3) "qq.com"
//按照分数由大到小查询 my_zset 集合的元素
127.0.0.1:6379> zrevrange my_zset 0 -1
1) "qq.com"
2) "taobao.com"
3) "baidu.com"
//查询元素'baidu.com'的分数值
127.0.0.1:6379> zscore my_zset 'baidu.com'
"1"
```

```
//查询元素'qq.com'的分数值
127.0.0.1:6379> zscore my_zset 'qq.com'
"3"
```

14.2.4　Spring Boot 整合 Redis

1. 创建 Spring Boot 项目

首先创建 Spring Boot Web 项目，添加 Redis 的依赖，并添加 Redis 的缓冲连接池，常用的缓冲连接池为 Lettuce 和 Jedis。Jedis 在多线程使用同一个连接时，线程是不安全的，所以要使用连接池，得为每个 Jedis 实例分配一个连接。Lettuce 在多线程使用同一连接实例时，线程是安全的。Redis 在默认情况下使用 Lettuce 连接池，本节引入 Lettuce，依赖如下：

```xml
<!-- Redis 相关依赖 -->
<dependency>
    <groupId>org.springframework.boot</groupId>
    <artifactId>spring-boot-starter-data-redis</artifactId>
</dependency>
<!-- Lettuce pool 缓冲连接池 -->
<dependency>
    <groupId>org.apache.commons</groupId>
    <artifactId>commons-pool2</artifactId>
</dependency>
```

2. 配置文件

在 application.properties 配置文件中，添加如下配置信息：

```
#Redis 数据库索引(默认为0)
spring.redis.database=0
#Redis 服务器地址
spring.redis.host=localhost
#Redis 服务器连接端口
spring.redis.port=6379
#Redis 服务器连接密码(默认为空)
spring.redis.password=
#连接池最大连接数(使用负值表示没有限制)
spring.redis.lettuce.pool.max-active=8
#连接池最大阻塞等待时间(使用负值表示没有限制)
spring.redis.lettuce.pool.max-wait=-1ms
#连接池中的最大空闲连接
spring.redis.lettuce.pool.max-idle=8
#连接池中的最小空闲连接
spring.redis.pool.min-idle=0
```

3. 创建实体类

创建一个 City 类，代码如下：

```java
public class City implements Serializable {
    private int id;
    private String name;
    private String country;
    //省略 get/set
    //toString()方法
}
```

4. 创建 Controller

实体类及 Redis 的连接信息添加完成后，创建 CityControlle 进行测试，代码如下：

```java
@RestController
public class CityController {
    @Autowired
    private RedisTemplate redisTemplate;
    @Autowired
    private StringRedisTemplate stringRedisTemplate;
    @GetMapping("/")
    public void testRedis(){
        ValueOperations<String, String> ops = stringRedisTemplate.opsForValue();
        //添加字符串
        ops.set("name", "beixi");
        String name = ops.get("name");
        System.out.println(name);
        ValueOperations opsForValue = redisTemplate.opsForValue();
        City city = new City(1, "北京", "中国");
        //添加实体类
        opsForValue.set("city", city);
        Boolean exists = redisTemplate.hasKey("city");
        System.out.println("redis 是否存在相应的 key: " + exists);
        //删除
        redisTemplate.delete("city");
        //更新
        redisTemplate.opsForValue().set("city", new City(2, "山西","中国"));
        //查询
        City c2 = (City) redisTemplate.opsForValue().get("city");
        System.out.println("从 redis 数据库中获取 city: " + c2.toString());
    }
}
```

RedisTemplate 和 StringRedisTemplate 都是 Spring Data Redis 为我们提供的模板类，用来对数据进行操作，都通过 Spring 提供的 Serializer 序列化到数据库。其中 StringRedisTemplate

是 RedisTemplate 的子类，只针对键值都是字符串的数据进行操作，采用的序列化方案是 StringRedisSerializer，而 RedisTemplate 可以操作对象，采用的序列化方案是 JdkSerializationRedisSerializer。在 Spring Boot 中默认提供这两个模板类，StringRedisTemplate 和 RedisTemplate 都提供了 Redis 的基本操作方法。

当然 RedisTemplate 和 StringRedisTemplate 还为我们提供了下面几个数据访问方法。

（1）opsForList：操作 list 数据。

（2）opsForSet：操作 set 数据。

（3）opsForZSet：操作 ZSet 数据。

（4）opsForHash：操作 hash 数据。

5．测试

在浏览器中输入 http://localhost:8080/访问，观察控制台打印的日志信息，如图 14-7 所示。当然我们还可以使用 RedisClient 客户端工具查看 Redis 缓冲数据库中的数据。

```
beixi
redis是否存在相应的key: true
从redis数据库中获取city: City{id=2, name='山西', country='中国'}
```

图 14-7　Redis 数据存储信息

14.2.5　Redis 缓冲在 Spring Boot 项目中的应用

在开发中，如果以相同的查询条件频繁查询数据库，会给数据库带来很大的压力。因此，我们需要对查询出来的数据进行缓存，这样客户端只需从数据库查询一次数据，然后放入缓存中，以后再次查询时可以从缓存中读取，这样便提高了数据的访问速度。开发步骤如下。

1．准备工作

开启 Redis 数据库，创建 Spring Boot Web 项目。使用 MySQL 数据库创建 user 表用于项目开发数据的存储，如图 14-8 所示。

id	name	pwd
1	beixi	123
2	jzj	123

图 14-8　user 表

2．导入依赖

在 pom.xml 文件中导入 Redis 数据库、持久层 MyBatis、MySQL 驱动等相关依赖，代码如下：

```xml
<!-- MyBatis 与 Spring Boot 2.x 的整合包 -->
<dependency>
    <groupId>org.mybatis.spring.boot</groupId>
    <artifactId>mybatis-spring-boot-starter</artifactId>
    <version>1.3.2</version>
</dependency>
```

```xml
<!-- MySQL JDBC 驱动 -->
<dependency>
    <groupId>mysql</groupId>
    <artifactId>mysql-connector-java</artifactId>
    <version>5.1.39</version>
</dependency>
<dependency>
    <groupId>org.springframework.boot</groupId>
    <artifactId>spring-boot-starter-data-redis</artifactId>
</dependency>
<!-- jedis 连接池 -->
<dependency>
    <groupId>redis.clients</groupId>
    <artifactId>jedis</artifactId>
</dependency>
<dependency>
    <groupId>org.springframework.boot</groupId>
    <artifactId>spring-boot-starter-cache</artifactId>
</dependency>
```

3. 配置文件

接下来在 application.yml 中配置 Redis 连接信息、MySQL 数据库连接等相关信息，代码如下：

```yaml
server:
  port: 8081
# 数据库连接
spring:
  datasource:
    url: jdbc:mysql://localhost:3306/test?useUnicode=true
    driver-class-name: com.mysql.jdbc.Driver
    username: root
    password: root
  ## Redis 配置
  redis:
    ## Redis 数据库索引(默认为 0)
    database: 0
    ## Redis 服务器地址
    host: localhost
    ## Redis 服务器连接端口
    port: 6379
    ## Redis 服务器连接密码(默认为空)
    password:
    jedis:
```

```yaml
    pool:
      ##连接池最大连接数(使用负值表示没有限制)
      #spring.redis.pool.max-active=8
      max-active: 8
      ##连接池最大阻塞等待时间(使用负值表示没有限制)
      #spring.redis.pool.max-wait=-1
      max-wait: -1
      ##连接池中的最大空闲连接
      #spring.redis.pool.max-idle=8
      max-idle: 8
      ##连接池中的最小空闲连接
      #spring.redis.pool.min-idle=0
      min-idle: 0
  ##连接超时时间(ms)
    timeout: 1200
#将Thymeleaf的默认缓存禁用,热加载生效
    thymeleaf:
      cache: false
#MyBatis的下画线转驼峰配置
configuration:
      map-underscore-to-camel-case: true
              #另外一种打印语句的方式
      log-impl: org.apache.ibatis.logging.stdout.StdOutImpl
    #打印SQL时的语句
logging:
  level:
    com:
      acong:
        dao: debug
```

4. 实体类

创建与数据库相对应的 User 实体类,代码如下:

```java
public class User implements Serializable {
    private int id;
    private String name;
    private String pwd;
    //省略get/set
    //省略有参/无参构造
}
```

5. 创建持久层

接着是 Mapper 持久层 Dao,这里主要用注解写比较方便,也可以使用 MyBatis 的 XML 配置文件写 SQL 语句,代码如下:

```
@Mapper
public interface UserDao {
    @Select("select * from user")
    List<User> queryAll();
    @Select("select * from user where id = #{id}")
    User findUserById(int id);
}
```

6．业务层

创建 UserService，这里主要使用 Redis 模板来写，代码如下：

```
@Service
public class UserService {
    @Autowired
    private UserDao userDao;
    @Autowired
    private RedisTemplate redisTemplate;
    public List<User> queryAll() {
        return userDao.queryAll();
    }
    /**
     * 获取用户策略：先从缓存中获取用户,没有用户信息则取数据表中的数据,再将数据写入缓存
     */
    public User findUserById(int id) {
        String key = "user_" + id;
        ValueOperations<String, User> operations = redisTemplate.opsForValue();
        //判断 Redis 中是否有键为 key 的缓存
        boolean hasKey = redisTemplate.hasKey(key);
        if (hasKey) {
            User user = operations.get(key);
            System.out.println("从缓存中获得数据：" + user.getName());
            System.out.println("------------------------------------");
            return user;
        } else {
            User user = userDao.findUserById(id);
            System.out.println("查询数据库获得数据：" + user.getName());
            System.out.println("------------------------------------");
            // 写入缓存
            operations.set(key, user, 5, TimeUnit.HOURS);
            return user;
        }
    }
}
```

7. 控制层

创建 UserController，用于暴露接口访问，代码如下：

```java
@RestController
public class UserController {
    @Autowired
    private UserService userService;
    @RequestMapping("/queryAll")
    public List<User> queryAll(){
        List<User> lists = userService.queryAll();
        return lists;
    }
    @RequestMapping("/findUserById")
    public Map<String, Object> findUserById(@RequestParam int id){
        User user = userService.findUserById(id);
        Map<String, Object> result = new HashMap<>();
        result.put("id", user.getId());
        result.put("name", user.getName());
        result.put("pwd", user.getPwd());
        return result;
    }
}
```

8. 测试

这里主要使用 RedisTemplate 来对 Redis 操作，每次访问 controller 暴露的接口，首先判断 Redis 缓存中是否存在该数据，若不存在就从数据库中读取数据，然后保存到 Redis 缓存中，当下次访问的时候，就直接从缓存中取出来。这样就不用每次都执行 SQL 语句，这样能够提高访问速度。但是在保存数据到缓存中时，需要设置键和值及超时删除，注意设置超时删除缓存时间不要太长，否则会给服务器带来压力。

图 14-9 控制台打印信息

启动 Spring Boot 项目，在浏览器中输入：http://localhost:8081/findUserById?id=1，当我们第一次访问数据时从数据库中获取，当再次访问时则从缓存中获取保存的数据，如图 14-9 所示。

14.3 整合 MongoDB

14.3.1 MongoDB 简介

MongoDB 是一个基于分布式文件存储的数据库，由 C++ 语言编写。旨在为 Web 应用提供可扩展的高性能数据存储解决方案。

MongoDB 是一个跨平台的面向文档的数据库，是当前 NoSQL 数据库产品中最热门的一种。它介于关系数据库和非关系数据库之间，是非关系数据库当中功能最丰富，最像关系数据库的产品。它支持的数据结构非常松散，类似于 JSON 的 BSON 格式，因此可以存储比较复杂的数据类型。Mongo 最大的特点是它支持的查询语言非常强大，其语法有点类似于面向对象的查询语言，几乎可以实现类似关系数据库单表查询的绝大部分功能，而且还支持对数据建立索引。它的特点是高性能、易部署、易使用，存储数据非常方便。MongoDB 有很多优点，但缺点也很明显，例如不能建立实体关系、没有事务管理机制等。

14.3.2　MongoDB 安装

MongoDB 提供了 Linux、Windows、Mac OS X 等操作系统的安装包。本书主要讲解 Windows 操作系统版本的 MongoDB，安装步骤如下。

（1）进入官网 https://www.mongodb.com/download-center/community，下载 MongoDB 服务端安装包，如图 14-10 所示。

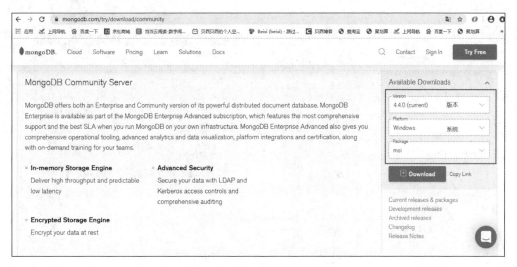

图 14-10　MongoDB 官方网站

（2）下载完成后，双击安装文件开始安装，然后根据提示单击 Next 按钮即可，当进入如图 14-11 所示这个界面的时候，去掉 Install MongoDB Compass 前的√，否则安装会变得特别慢。

（3）安装完成后，在所安装目录的 data 目录下，创建一个 db 文件夹（例如 F:\Program Files\MongoDB\data\db），用于数据库的存放。因为启动 MongoDB 服务之前必须创建数据库存放的文件夹，否则命令不会自动创建，而且不能成功启动。

（4）配置环境变量（MongoDB 安装目录下的 bin 目录），如图 14-12 所示。

（5）打开命令窗口，输入：mongod --dbpath F:\Program Files\MongoDB\data\db（这是自定义安装目录）命令启动 MongoDB 服务，如图 14-13 所示。

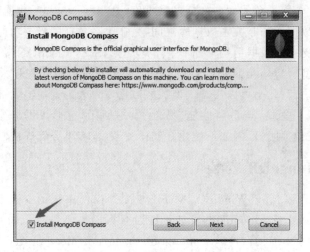

图 14-11　安装 MongoDB

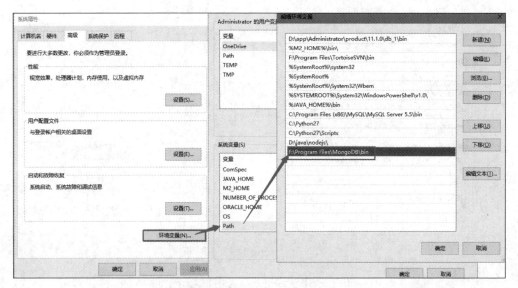

图 14-12　环境变量配置

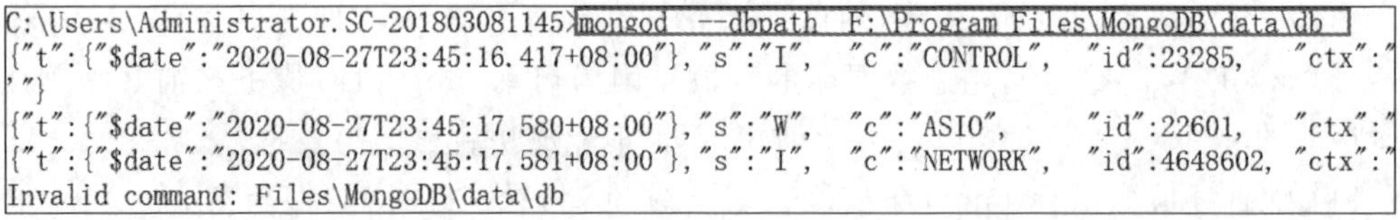

图 14-13　启动 MongoDB

(6) 启动 MongoDB 客户端，另外打开一个 cmd 命令窗口，执行 mongo 命令，出现如图 14-14 所示界面，说明已经启动成功了。

```
C:\Users\Administrator.SC-201803081145>mongo
MongoDB shell version v4.4.0
connecting to: mongodb://127.0.0.1:27017/?compressors=disabled&gssa
Implicit session: session { "id" : UUID("244a66b0-e351-4f3f-8636-4b
MongoDB server version: 4.4.0
```

图 14-14　启动 MongoDB 客户端

当然一个好的开发工具是开发过程中的一个重要组成部分。MongoDB shell 用来管理操作工作很方便，但在处理大量数据时，用户界面变得相当重要，能够为我们带来更便利的操作。连接 MongoDB 数据库的方式很多，可以使用功能强大的 NoSQL Manager for MongoDB 可视化客户端来连接 MongoDB 数据库。下载地址：https://www.mongodbmanager.com/download，下载完成后，安装及操作比较简单，这里省略安装步骤。

14.3.3　常用命令

MongoDB 常用的一些 SQL 语句如下。

1. 查询

(1) show dbs；查看所有数据库。

(2) show tables；查询表。

(3) db.user.find()；查询 user 集合(表)。

(4) db.user.findOne()；查询第一条数据。

(5) db.user.find({"age":{$gt:10}})；根据条件查询。

(6) show collections；显示所有集合。

(7) db.user.find({"age":{$gt:"10"}})年龄大于 10 的数据。

2. 切换数据库

使用 use test。

3. 增加

(1) db.createCollection("test")创建集合(创建数据库)。

(2) db.user.insert({"id":"1","name":"beixi","age":"18"}) 直接创建表及插入数据。

4. 修改

db.user.update({"id":"1"},{$set:{"name":"jzj"}})。

5. 删除

(1) db.dropDatabase() 要先切换到要删除的数据库，然后执行该语句。

(2) db.user.drop() 删除集合/表的 user。

（3）db.user.remove({"name":"jzj"}) 删除某一条数据。

（4）db.user.remove({}) 删除表内所有数据。

关于 MongoDB 更多的 SQL 语句大家可以参考官网或者查阅相关资料进行学习，下面我们来进行具体的整合过程。

14.3.4 Spring Boot 整合 MongoDB

本节只是实现 Spring Boot 整合 MongoDB 的一个 demo，所以没有写 service 层，相关操作都写到了 Dao 层。

1. 创建项目

首先创建 Spring Boot Web 项目，并添加 MongoDB 依赖，代码如下：

```xml
<dependency>
    <groupId>org.springframework.boot</groupId>
    <artifactId>spring-boot-starter-data-mongodb</artifactId>
</dependency>
```

2. 配置文件

在 application.yml 中配置 MongoDB 连接信息，代码如下：

```yaml
spring:
  data:
    mongodb:
      uri: mongodb://localhost:27017/test
```

其中 27017 是端口，test 是数据库。

MongoDB 数据库在默认情况下没有用户名及密码，不用安全验证，只要连接上服务就可以进行 CRUD 操作。

3. 创建实体类

在实际项目中，我们使用最多的还是对象，在 MongoDB 里，所有的对象都被认为是文档，对象的属性对 MongoDB 来说就是一个个字段，所以最方便的还是对对象文档的操作，这里模拟一个对象，代码如下：

```java
public class Book {
    private Integer id;
    private String name;
    private Integer price;
    private Date updateTime;
    //省略 get/set/构造方法
}
```

4. 持久层

操作 Mongo 数据及具体业务类，代码如下：

```java
@Component
public class MongoTestDao {
    @Autowired
    private MongoTemplate mongoTemplate;
    /*保存对象*/
    public String saveObj(Book book) {
        book.setUpdateTime(new Date());
        mongoTemplate.save(book);
        return "添加成功";
    }
    /*查询所有*/
    public List<Book> findAll() {
        return mongoTemplate.findAll(Book.class);
    }
    /*根据名称查询*/
    public Book getBookByName(String name) {
        Query query = new Query(Criteria.where("name").is(name));
        return mongoTemplate.findOne(query, Book.class);
    }
    /*更新对象*/
    public String updateBook(Book book) {
        Query query = new Query(Criteria.where("_id").is(book.getId()));
        Update update = new Update().set("name", book.getName()).set("updateTime",
                new Date());
        //updateFirst 更新查询返回结果集的第一条
        Template.updateFirst(query, update, Book.class);
        //updateMulti 更新查询返回结果集的全部
        //mongoTemplate.updateMulti(query,update,Book.class);
        //upsert 更新对象不存在则添加
        //mongoTemplate.upsert(query,update,Book.class);
        return "success";
    }
    /* 删除对象 */
    public String deleteBook(int id) {
        Query query = new Query(Criteria.where("id").is(id));
        mongoTemplate.remove(query, Book.class);
        return "success";
    }
}
```

5. 控制层

创建控制层 MongoTestController 类，代码如下：

```java
@RestController
public class MongoTestController {
    @Autowired
    private MongoTestDao mongoTestDao;
    @PostMapping("/mongo/save")
    public String saveObj(@RequestBody Book book) {
        return mongoTestDao.saveObj(book);
    }
    @GetMapping("/mongo/findAll")
    public List<Book> findAll() {
        return mongoTestDao.findAll();
    }
    @GetMapping("/mongo/findOneByName")
    public Book findOneByName(@RequestParam String name) {
        return mongoTestDao.getBookByName(name);
    }
    @PostMapping("/mongo/update")
    public String update(@RequestBody Book book) {
        return mongoTestDao.updateBook(book);
    }
    @PostMapping("/mongo/delOne")
    public String delOne(int id) {
        return mongoTestDao.deleteBook(id);
    }
}
```

6. 测试

启动项目，使用 Postman 进行测试。

调用 controller 中的新增方法，先在 MongoDB 数据库中保存一条数据，如图 14-15 所示。

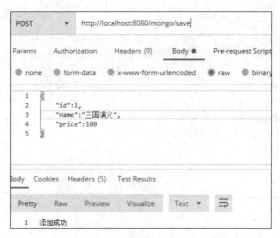

图 14-15 新增一条数据

在命令窗口中查看数据信息,如图14-16所示。

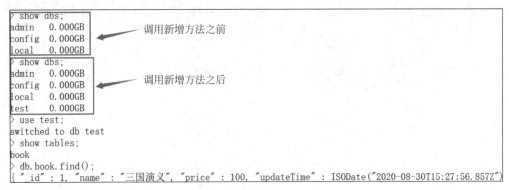

图14-16 命令窗口中查看新增数据信息

我们发现,在发送请求之前使用show dbs是看不到test库的。当我们发送请求之后,就可以看到test库了,同时test库中也有了book表,使用db.book.find()命令可以看到数据。

再测试一下查询所有数据信息,之前笔者已经保存了两条数据,如图14-17所示。

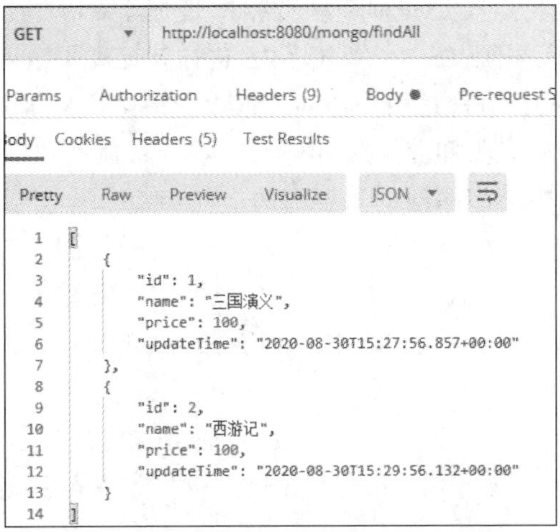

图14-17 查询所有数据信息

测试根据条件查询,如图14-18所示。

其他的测试就不一一演示了。由于MongoDB独特的数据处理方式,可以将热点数据加载到内存,故而对查询来讲,会非常快(当然也会非常消耗内存),同时由于采用了BSON的方式存储数据,故而对JSON格式数据具有非常好的支持性及友好的表结构修改性,文档式的存储方式使数据友好可见。数据库的分片集群负载具有非常好的扩展性及非常不错的自动故障转移。

图 14-18　根据 name 条件查询

14.4　整合 MySQL

14.4.1　MySQL 简介

MySQL 是一个关系型数据库管理系统，由瑞典 MySQL AB 公司开发，目前属于 Oracle 公司。MySQL 也可以说是一种关联数据库管理系统，关联数据库将数据保存在不同的表中，而不是将所有数据放在一个大仓库内，这样就提高了数据提取速度并提高了灵活性。MySQL 所使用的 SQL 语言是用于访问数据库的最常用标准化语言。MySQL 软件采用了双授权政策，它分为社区版和商业版，由于其体积小、速度快、总体拥有成本低，尤其是开放源码这一特点，一般中小型网站的开发都选择 MySQL 作为网站数据库。

与其他的大型数据库例如 Oracle、DB2、SQL Server 等相比，MySQL 自有它的不足之处，但是这丝毫也没有减少它受欢迎的程度。对于个人使用者和中小型企业来说，MySQL 提供的功能已经绰绰有余，而且由于 MySQL 是开放源码软件，因此可以大大降低总体拥有成本。

MySQL 的安装很简单，大家可以到官方网址 https://dev.mysql.com/downloads/mysql/下载并安装 MySQL 服务端。关于 MySQL 这里就不再赘述。

14.4.2　Spring Boot 整合 MySQL

Spring Boot 是目前 Java 世界最流行的一个企业级解决方案框架。它深度绑定了依赖注入和面向切片两种编程思想，并且通过自动化的方式减少了开发过程中的大量烦琐的配置和通用的配置型编码，让编程人员可以更加聚焦于业务，解决实际的问题。因此 Spring Boot 集成 MySQL 非常简单。Spring Boot 在连接数据库时有 4 种方式：

（1）采用 JDBC 直接连接。

（2）采用 JdbcTemplate 连接。

（3）采用 SpringDataJPA 连接。

(4) 框架连接。

采用 JDBC 方式直接连接烦琐，易错，我们直接略过，不做考虑。通过 MyBatis、SpringDataJPA 等连接，我们后续再讲。JdbcTemplate 在 JDBC 的基础上做了大量的封装，本节我们采用 JdbcTemplate 连接 MySQL。

1. 引入依赖

在 pom.xml 文件中引入 MySQL 驱动及 JDBC 依赖，代码如下：

```xml
<!-- MySQL 连接 Java 的驱动程序 -->
<dependency>
    <groupId>mysql</groupId>
    <artifactId>mysql-connector-java</artifactId>
</dependency>
<!-- 支持通过 JDBC 连接数据库 -->
<dependency>
    <groupId>org.springframework.boot</groupId>
    <artifactId>spring-boot-starter-jdbc</artifactId>
</dependency>
```

2. 添加数据库配置

在 application.yml 文件中添加如下的配置：

```yml
spring:
  datasource:
    #MySQL 连接信息 serverTimezone=GMT%2B8 解决时区时间差报错问题
    url: jdbc:mysql://localhost:3306/test?serverTimezone=GMT%2B8
    #账号
    username: root
    #密码
    password: root
    #驱动
    driver-class-name: com.mysql.jdbc.Driver
```

3. 设计表和实体

配置完信息之后，在 test 数据库中新建一张用户表 user，具体 SQL 语句如下：

```sql
-- ----------------------------
-- Table structure for user
-- ----------------------------
DROP TABLE IF EXISTS 'user';
CREATE TABLE 'user' (
  'id' int NOT NULL AUTO_INCREMENT COMMENT '主键',
  'name' varchar(10) DEFAULT '' COMMENT '用户姓名',
  'password' varchar(32) DEFAULT '' COMMENT '密码',
```

```
    PRIMARY KEY ('id')
);

INSERT INTO 'user' VALUES (1, 'beixi', '123456');
INSERT INTO 'user' VALUES (2, 'jzj', '123456');
```

表和数据准备好之后,在项目中新建 User 实体类,代码如下:

```
public class User {
    private int id;
    private String name;
    private String password;
//省略 get/set 方法 toString()方法
}
```

4. 控制层

创建 UserController 类,代码如下:

```
@Controller
public class UserController {
    @Autowired
    JdbcTemplate jdbcTemplate;

    @ResponseBody
    @RequestMapping("/list")
    public List mySqlTest() {
        String sql = "select * from user";
        /*query()是 JdbcTemplate 对象中的方法,RowMapper 对象可以查询数据库中的数据*/
        List<User> users = jdbcTemplate.query(sql, new RowMapper<User>() {
            @Override
            /*RowMapper 对象通过调用 mapRow()方法将数据库中的每一行数据封装成 User 对象,并返回*/
            public User mapRow(ResultSet rs, int i) throws SQLException {
                User user = new User();
                user.setId(rs.getInt("id"));
                user.setName(rs.getString("name"));
                user.setPassword(rs.getString("password"));
                return user;
            }
        });
        System.out.println("查询成功:" + users);
        return users;
    }
}
```

代码解释如下。

（1）JdbcTemplate：JDBC 连接数据库的工具类。

（2）Query()：query()是 JdbcTemplate 对象中的方法，RowMapper 对象可以查询数据库中的数据。

（3）mapRow()：RowMapper 对象通过调用 mapRow()方法将数据库中的每一行数据封装成 User 对象，并返回。

访问 http://localhost:8080/list 路径，即可查询数据信息，如图 14-19 所示。

```
查询成功: [User{id=1, name='beixi', password='123456'}, User{id=2, name='jzj', password='123456'}]
```

图 14-19　查询的数据信息

第 15 章 Spring Boot 整合持久层技术

持久层的实现是和数据库紧密相连的。在 Java 领域内,访问数据库的通常做法是使用 JDBC。JDBC 使用灵活而且访问速度较快,但 JDBC 不仅需要操作对象,还需要操作关系,并不是完全的面向对象编程。近年来出现了许多持久层框架,这些框架为持久层的实现提供了更多选择。目前主流的持久层框架包括:MyBatis、JPA 等。这些框架都对 JDBC 进行了封装,使得业务逻辑开发人员不再面对关系型操作,简化了持久层的开发。

15.1 整合 JdbcTemplate

Spring JDBC 抽象框架 core 包提供了 JDBC 模板类,其中 JdbcTemplate 是 core 包的核心类,所以其他模板类都是基于它封装完成的,JDBC 模板类是第一种工作模式。

JdbcTemplate 类通过模板设计模式帮助我们消除了冗长的代码,只做需要做的事情(即可变部分),并且帮我们做了固定部分,如连接的创建及关闭等。

Spring Boot 自动配置了 JdbcTemplate 数据模板。在 14.4.2 节中已经引入了 JdbcTemplate 模板,本节在其基础上进行开发。

1. 配置文件

在 application.yml 文件中添加如下配置信息:

```
spring:
  datasource:
    url: jdbc:mysql://localhost:3306/test?serverTimezone=GMT%2B8
    username: root
    password: root
    driver-class-name: com.mysql.jdbc.Driver
```

JDBC 默认采用 org.apache.tomcat.jdbc.pool.DataSource 数据源,数据源的相关配置参考都封装在 org.springframework.boot.autoconfigure.jdbc.DataSourceConfiguration 类中。默认使用 Tomcat 连接池,可以使用 spring.datasource.type 指定自定义的数据源类型。

Spring Boot 默认支持的数据源：

（1）org.apache.tomcat.jdbc.pool.DataSource。

（2）HikariDataSource。

（3）BasicDataSource。

默认采用 org.apache.tomcat.jdbc.pool.DataSource，但是实际开发中很少使用这个数据源，可以使用 spring.datasource.type 指定自定义的数据源类型，例如 Druid、c3p0。

2. 整合 Druid 数据源

导入 Druid 依赖，代码如下：

```xml
<dependency>
    <groupId>com.alibaba</groupId>
    <artifactId>druid</artifactId>
    <version>1.1.10</version>
</dependency>
```

修改 application.yml 配置文件，代码如下：

```yaml
spring:
  datasource:
    url: jdbc:mysql://localhost:3306/test?serverTimezone=GMT%2B8
    username: root
    password: root
    driver-class-name: com.mysql.jdbc.Driver
    type: com.alibaba.druid.pool.DruidDataSource
    #数据源其他配置信息
    initialSize: 5
    minIdle: 5
    maxActive: 20
    maxWait: 60000
    timeBetweenEvictionRunsMillis: 60000
    minEvictableIdleTimeMillis: 300000
    validationQuery: SELECT 1 FROM DUAL
    testWhileIdle: true
    testOnBorrow: false
    testOnReturn: false
    poolPreparedStatements: true
    filter:
      stat:
        enabled: true
        log-slow-sql: true
      wall:
        enabled: true
```

配置 type：com.alibaba.druid.pool.DruidDataSource 来切换数据源，但是文件中数据

源其他配置信息现在还不会起任何作用。

因为 Druid 是第三方数据源，在 Spring Boot 中并没有自动配置类，需要自己配置数据。所以我们需要自定义一个 Druid 数据源进行绑定，代码如下：

```
@Configuration
public class Config {
    /*把配置文件中 spring.datasource.initialSize 等属性和 DruidDataSource 中属性进行绑定*/
    @ConfigurationProperties(prefix = "spring.datasource")
    @Bean
    public DruidDataSource druidDataSource () {
        return new DruidDataSource();
    }
}
```

配置 Druid 的监控和过滤器，代码如下：

```
@Configuration
public class DuridConfig {
    /**
     * 配置 Druid 的监控
     * 1、配置一个管理后台的 Servlet,拦截登录
     * @return
     */
    @Bean
    public ServletRegistrationBean statViewServlet(){
        ServletRegistrationBean servletRegistrationBean = new ServletRegistrationBean(new StatViewServlet(),"/druid/*");
        Map<String,String> initParams = new HashMap<>();
        initParams.put("loginUsername","admin");
        initParams.put("loginPassword","123456");
        initParams.put("allow","");  // 允许所有访问
        servletRegistrationBean.setInitParameters(initParams);
        return servletRegistrationBean;
    }
    // 配置一个 Web 监控的 filter,哪些请求会被监控,哪些排除
    @Bean
    public FilterRegistrationBean webStatFilter() {
        FilterRegistrationBean bean = new FilterRegistrationBean(new WebStatFilter());
        Map<String,String> initParams = new HashMap<>();
        initParams.put("exclusions","*.js,*.css,/druid/*");
        bean.setInitParameters(initParams);
        bean.setUrlPatterns(Arrays.asList("/*"));
        return bean;
    }
}
```

代码解释如下。

（1）@Configuration：@Configuration 用于定义配置类，可替换 XML 配置文件，被注解的类内部包含一个或多个 @Bean 注解方法。可以被 AnnotationConfigApplicationContext 或者 AnnotationConfigWebApplicationContext 进行扫描。用于构建 Bean 定义及初始化 Spring 容器。

（2）@Bean：等同于 XML 配置文件中的 <bean> 配置。

启动项目，当访问 http://localhost:8080/druid 时会进入监控的登录页面，如图 15-1 所示。在登录成功之后，可以对数据源、SQL、Web 应用等进行监控，如图 15-2 所示。

图 15-1　Druid 登录页面

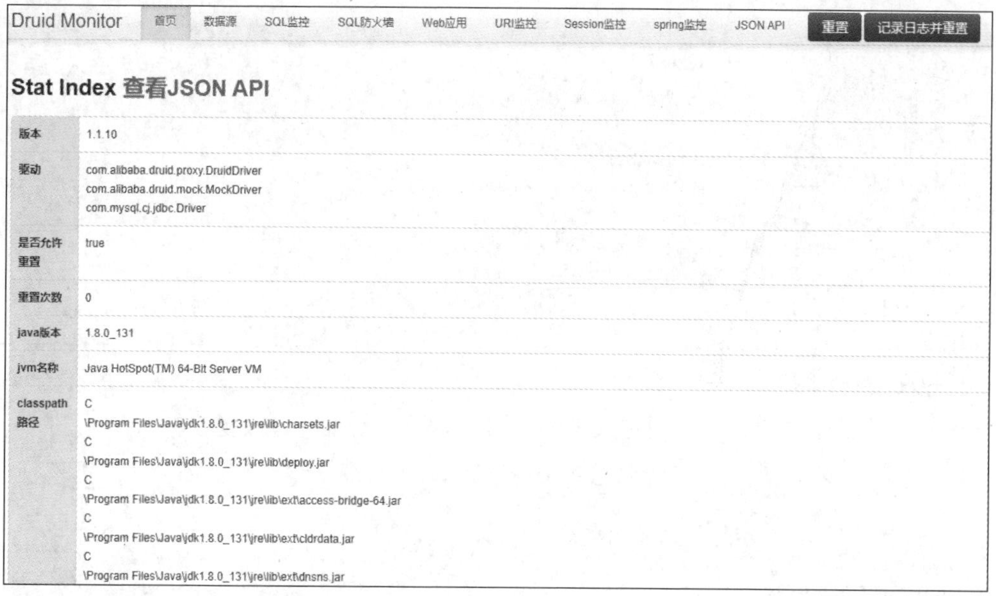

图 15-2　Druid 监控页面

15.2　整合 MyBatis

15.2.1　MyBatis 简介

MyBatis 的前身叫 iBatis，本来是 Apache 的一个开源项目，2010 年这个项目由 Apache Software Foundation 迁移到了 Google Code，并且改名为 MyBatis。MyBatis 支持普通 SQL 查询，其存储过程和高级映射是优秀的持久层框架。MyBatis 消除了绝大多数的 JDBC 代码和参数的手工设置及结果集的检索。MyBatis 使用简单的 XML 或注解用于配置和原始映射,将接口和 Java 的 POJOs(Plan Old Java Objects,普通的 Java 对象)映射成数据库中的记录。

15.2.2　Spring Boot 整合 MyBatis

15.2 节介绍了 Spring Boot 中最简单的数据持久化方案 JdbcTemplate,JdbcTemplate 虽然简单,但是用得并不多,因为它没有 MyBatis 方便,在 Spring+Spring MVC 中整合 MyBatis 步骤还是有点复杂,要配置多个 Bean,Spring Boot 中对此做了进一步简化,使 MyBatis 基本上可以做到开箱即用,本节讲解 Spring Boot 中 MyBatis 如何使用。

1. 创建项目

首先创建 Spring Boot Web 项目,添加 MyBatis、数据库驱动等依赖,代码如下:

```xml
<!-- MySQL 数据库驱动 -->
<dependency>
    <groupId>mysql</groupId>
    <artifactId>mysql-connector-java</artifactId>
</dependency>
<!-- MyBatis -->
<dependency>
    <groupId>org.mybatis.spring.boot</groupId>
    <artifactId>mybatis-spring-boot-starter</artifactId>
    <version>2.1.0</version>
</dependency>
```

2. 设计表和实体类

在 test 数据库中新建一张用户表 user,具体 SQL 语句如下:

```sql
-- ----------------------------
-- Table structure for user
-- ----------------------------
DROP TABLE IF EXISTS 'user';
CREATE TABLE 'user' (
  'id' int NOT NULL AUTO_INCREMENT COMMENT '主键',
  'name' varchar(10) DEFAULT '' COMMENT '用户姓名',
  'password' varchar(32) DEFAULT '' COMMENT '密码',
  PRIMARY KEY ('id')
);

INSERT INTO 'user' VALUES (1, 'beixi', '123456');
INSERT INTO 'user' VALUES (2, 'jzj', '123456');
```

表和数据准备好之后,在项目中新建 User 实体类,代码如下:

```
public class User {
    private int id;
    private String name;
    private String password;
//省略 get/set 方法 toString()方法
}
```

3. 配置文件

在 application.yml 中配置数据库及 MyBatis 的基本信息：

```yaml
#基本属性
spring:
  datasource:
    url: jdbc:mysql://localhost:3306/test?serverTimezone=GMT%2B8&characterEncoding=utf-8
    username: root
    password: root
    driver-class-name: com.mysql.jdbc.Driver
#MyBatis 的相关配置
mybatis:
  #Mapper 映射 XML 文件,建议写在 resources 目录下
  mapper-locations: classpath:mappers/*.xml
  #Mapper 接口存放的目录
  type-aliases-package: com.beixi.mapper
  #开启驼峰命名
  configuration:
    map-underscore-to-camel-case: true
```

4. 编写 Mapper 接口

配置完成后，MyBatis 就可以创建 Mapper 使用了，例如我们直接创建 UserMapper 接口，代码如下：

```java
//com.beixi.mapper.UserMapper
@Mapper//指定这是一个操作数据库的 Mapper
public interface UserMapper {
    List<User> findAll();
    int addUser(User user);
}
```

代码解释如下。

（1）@Mapper：接口类上需要使用@Mapper 注解,不然 Spring Boot 无法扫描到。

（2）有两种方式指明该类是一个 Mapper：一种方法是在上面的 Mapper 层对应的类上面添加 @Mapper 注解即可,但是这种方法有个弊端,当我们有很多个 Mapper 时,那么每

一个类上面都得添加 @Mapper 注解。另一种比较简便的方法是在 Spring Boot 启动类上添加 @MaperScan 注解,来扫描一个包下的所有 Mapper,如:@MapperScan("com.beixi.mapper"),表示扫描 com.beixi.mapper 包下所有为 Mapper 的接口。

5. Mapper 映射

使用原始的 XML 方式,纯注解式不利于复杂 SQL 的编写,需要新建 UserMapper.xml 文件,在上面的 application.yml 配置文件中,我们已经定义了 XML 文件的路径:classpath:mappers/*.xml,所以在 resources 目录下新建一个 mappers 文件夹,然后创建一个 UserMapper.xml 文件,代码如下:

```xml
<?xml version = "1.0" encoding = "UTF - 8" ?>
<!DOCTYPE mapper PUBLIC " - //mybatis.org//DTD Mapper 3.0//EN"" http://mybatis.org/dtd/mybatis - 3 - mapper.dtd">
<mapper namespace = "com.beixi.mapper.UserMapper">
    <resultMap id = "UserResult" type = "com.beixi.entity.User">
        <id column = "id" property = "id"></id>
        <result column = "name" property = "name"></result>
        <result column = "password" property = "password"></result>
    </resultMap>
    <select id = "findAll" resultMap = "UserResult">
        SELECT * FROM user
    </select>
    <insert id = "addUser" parameterType = "com.beixi.entity.User">
        insert into user (name,password) values (#{name},#{password});
    </insert>
</mapper>
```

注意:

(1) namespace 中需要与使用 @Mapper 的接口对应。

(2) UserMapper.xml 文件名称必须与使用 @Mapper 的接口一致。

(3) 标签中的 id 必须与 @Mapper 接口中的方法名一致,且参数一致。

6. 创建 Service 和 Controller

创建 UserService 和 UserController 类,代码如下:

```java
@Service
public class UserService {
    @Autowired
    private UserMapper userMapper;

    public List<User> findAll() {
        return userMapper.findAll();
    }
}
```

```
    public int addUser(User user){
        return userMapper.addUser(user);
    }
}

@RestController
@RequestMapping("/user")
public class UserController {
    @Autowired
    private UserService userService;
    @RequestMapping("/findAll")
    public List<User> findAll(){
        return userService.findAll();
    }
    @PostMapping("/add")
    public String add( User user){
        int i = userService.addUser(user);
        if (i>0) {return "success";}
        return "fail";
    }
}
```

7．测试

启动项目，在 Postman 中测试添加用户功能，如图 15-3 所示，以及查询所有的用户信息，如图 15-4 所示。

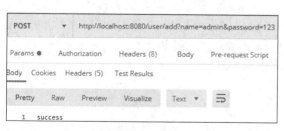

图 15-3　添加用户

图 15-4　查询用户

15.3　Spring Data JPA

我们知道Java持久层框架访问数据库的方式大致分为两种。一种以SQL核心，封装一定程度的JDBC操作，例如：MyBatis。另一种以Java实体类为核心，将实体类和数据库表之间建立映射关系，也就是我们所说的ORM框架，如：Hibernate、Spring Data JPA。本节我们就来了解一下什么是Spring Data JPA。

15.3.1　JPA、Spring Data、Spring Data JPA 的故事

1. 什么是JPA

在开始学习Spring Data JPA之前我们需要先了解一下什么是JPA，因为Spring Data JPA是建立的JPA的基础之上的。

JPA全称为Java Persistence API(Java持久层API)，它是Sun公司在Java EE 5中提出的Java持久化规范。它为Java开发人员提供了一种对象/关联映射工具，来管理Java应用中的关系数据，JPA吸取了目前Java持久化技术的优点，旨在规范、简化Java对象的持久化工作。很多ORM框架实现了JPA的规范，如：Hibernate、EclipseLink。也就是JPA统一了Java应用程序访问ORM框架的规范，而Hibernate是一个ORM框架，Hibernate是JPA的一种实现，是一个框架。

2. 什么是Spring Data

Spring Data是Spring社区的一个子项目，主要用于简化数据访问，其主要目标是使数据库的访问变得方便快捷。它不仅支持关系型数据库，也支持非关系型数据库。

3. 什么是Spring Data JPA

Spring Data JPA是在实现了JPA规范的基础上封装的一套JPA应用框架，虽然ORM框架都实现了JPA规范，但是在不同的ORM框架之间切换仍然需要编写不同的代码，而使用Spring Data JPA能够方便大家在不同的ORM框架之间进行切换而不需要更改代码。Spring Data JPA旨在通过将统一ORM框架的访问持久层的操作，来提高开发人员的效率。Spring Data JPA给我们提供的主要的类和接口，如图15-5所示。

总的来说，JPA是ORM规范，Hibernate、TopLink等是JPA规范的具体实现，这样的好处是开发者可以面向JPA规范进行持久层的开发，而底层的实现则是可以切换的。Spring Data JPA则是在JPA之上添加另一层抽象(Repository层的实现)，极大地简化持久层开发及ORM框架切换的成本，如图15-6所示。

15.3.2　整合 Spring Data JPA

本节主要介绍Spring Boot中集成Spring Data JPA、服务层类开发、如何通过Spring Data JPA快速实现基本的增、删、改、查，以及自定义查询方法等，具体步骤如下。

第15章 Spring Boot整合持久层技术

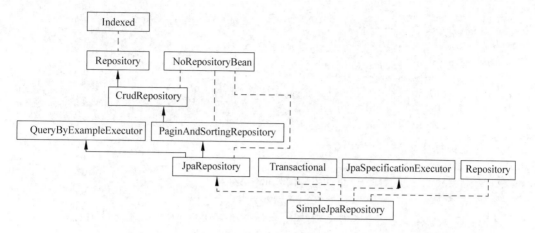

图 15-5　Spring Data JPA UML 类图

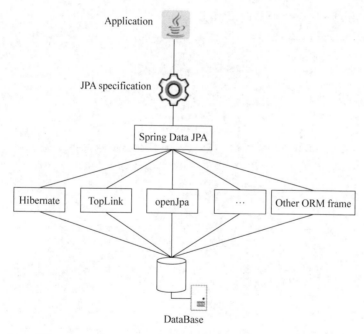

图 15-6　JPA、Hibernate、Spring Data JPA 三者之间的关系

1. 引入依赖

首先创建 Spring Boot Web 项目，添加 Spring Data JPA 的依赖，代码如下：

```xml
<!-- Spring Data JPA 依赖 -->
<dependency>
    <groupId>org.springframework.boot</groupId>
    <artifactId>spring-boot-starter-data-jpa</artifactId>
```

```xml
</dependency>
<!-- MySQL 驱动 -->
<dependency>
    <groupId>mysql</groupId>
    <artifactId>mysql-connector-java</artifactId>
</dependency>
```

2. 创建数据库

```sql
CREATE DATABASE 'jpa' DEFAULT CHARACTER SET utf8;
```

注意：只需创建数据库，不需要创建表。当启动项目后，JPA会帮助我们自动创建一个和实体类相对应的表。

3. 配置文件

在application.yml文件中添加如下配置：

```yaml
server:
  port: 8080
  servlet:
    context-path: /
spring:
  datasource:
    url: jdbc:mysql://localhost:3306/jpa?serverTimezone=GMT%2B8&characterEncoding=utf-8
    username: root
    password: root
  jpa:
    database: MySQL
    database-platform: org.hibernate.dialect.MySQL5InnoDBDialect
    show-sql: true           #日志中显示SQL语句
    hibernate:
        ddl-auto: update     #自动更新
```

ddl-auto 属性共有5个值，分别如下。

（1）create：每次运行程序时，都会重新创建表，故而数据会丢失。

（2）create-drop：每次运行程序时会先创建表结构，然后待程序结束时清空表。

（3）upadte：每次运行程序，如果没有表则创建表，如果对象发生改变则更新表结构，原有数据不会清空，只会更新（推荐使用）。

（4）validate：运行程序会校验数据与数据库的字段类型是否相同，字段不同会报错。

（5）none：禁用DDL处理。

4. 实体类

创建User实体类，代码如下：

```
@Entity                              //表示该类为实体类
@Table(name = "t_user")              //声明该类映射到数据库的表
public class User {
    @Id                              //代表主键
    @GeneratedValue(strategy = GenerationType.IDENTITY)          //主键自增的策略
    private Integer id;
    @Column(name = "username", unique = true, nullable = false, length = 64)
    private String username;
    @Column(name = "password", nullable = false, length = 64)
    private String password;
    @Column(name = "email", length = 64)
    private String email;
//省略 get/set
}
```

5. 创建 Dao 层

接着创建 UserRepository 接口,该接口只需继承 JpaRepository 接口。JpaRepository 中封装了基本的数据操作方法,有基本的增、删、改、查、分页、排序等,代码如下:

```
public interface UserRepository extends JpaRepository<User, Integer> {
}
```

6. Controller 层

这里为了结构简单省略 Service 层,创建 UserController 采用 Restful 风格实现增、删、改、查,代码如下:

```
@RestController
@RequestMapping("/users")
public class UserController {
    @Autowired
    private UserRepository userRepository;
    @PostMapping()                   //添加
    public User saveUser(@RequestBody User user) {
        return userRepository.save(user);
    }
    @DeleteMapping("/{id}")          //删除
    public void deleteUser(@PathVariable("id") int id) {
        userRepository.deleteById(id);
    }
    @PutMapping("/{id}")
    public User updateUser(@PathVariable("id") int userId, @RequestBody User user) {
        user.setId(userId);
        return userRepository.saveAndFlush(user);
```

```
    }
    @GetMapping("/{id}")
    public User getUserInfo(@PathVariable("id") int userId) {
        Optional<User> optional = userRepository.findById(userId);
        return optional.orElseGet(User::new);
    }
    @GetMapping("/list")
    public Page<User> pageQuery(@RequestParam(value = "pageNum", defaultValue = "1")
        Integer pageNum, @RequestParam(value = "pageSize", defaultValue = "10")
        Integer pageSize) {
        return userRepository.findAll(PageRequest.of(pageNum - 1, pageSize));
    }
}
```

代码解释：

上述类中使用的 save()、deleteById() 等方法是由 JpaRepository 接口提供的，可以很方便地直接使用。

7. 测试

启动程序，查询数据库就可以看到，JPA 已经自动帮我们创建了表，如图 15-7 所示。

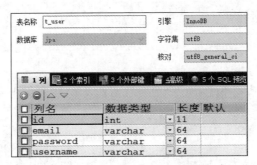

图 15-7　JPA 自动创建的实体类映射表

接下来使用 Postman 测试增加接口，如图 15-8 所示。

其余的增、删、改、查接口就不再给大家一一测试了，大家自行测试即可。

8. 自定义查询方法

我们除了使用 JpaRepository 接口提供的方法之外，还可以自定义查询方法。如我们要根据 username 和 password 查询 User 实例，可在 UserController 中添加一个 Rest 接口，代码如下：

```
@PostMapping("/login")
public User getPerson( String username,String password) {
    return userRepository.findByUsernameAndPassword(username, password);
}
```

图 15-8 调用添加接口

并在 UserRepository 接口中添加如下查询方法：

```
/*相当于 select * from t_user where username = ? and password = ? */
User findByUsernameAndPassword(String username, String password);
```

重启项目，在 Postman 中进行测试，结果如图 15-9 所示。

图 15-9 自定义查询方法测试

我们也可以在日志中看到 Hibernate 输出的 SQL 日志：

```
Hibernate: select user0_.id as id1_0_, user0_.email as email2_0_, user0_.password as password3_0_, user0_.username as username4_0_ from t_user user0_ where user0_.username = ? and user0_.password = ?
```

这是因为在 Spring Data JPA 中，只要自定义的方法符合既定规范，Spring Data 就会分析开发者的意图，从而根据方法名生成相应的 SQL 查询语句。

JPA 在这里遵循 Convention over configuration(约定大约配置)的原则,遵循 Spring 及 JPQL 定义的方法命名。Spring 提供了一套可以通过命名规则进行查询构建的机制。这套机制首先通过方法名过滤一些关键字,例如 find…By、read…By、query…By、count…By 和 get…By。系统会根据关键字将命名解析成两个子语句,第一个 By 是区分这两个子语句的关键词。这个 By 之前的子语句是查询子语句(指明返回要查询的对象),后面的部分是条件子语句。如果直接不区分就是 find…By 返回的就是定义 Respository 时指定的领域对象集合,同时 JPQL 中也定义了丰富的关键字:And、Or、Between 等,Spring Data JPA 支持的方法命名规则如表 15-1 所示。

表 15-1 命名规则

关键字	方法名举例	对应的 SQL
And	findByLastnameAndFirstname	…where x.lastname=?1 and x.firstname=?2
Or	findByLastnameOrFirstname	…where x.lastname=?1 or x.firstname=?2
Is,Equals	findByFirstname,findByFirstnameIs,findByFirstnameEquals	…where x.firstname=?1
Between	findByStartDateBetween	…where x.startDate between ?1 and ?2
LessThan	findByAgeLessThan	…where x.age<?1
LessThanEqual	findByAgeLessThanEqual	…where x.age<=?1
GreaterThan	findByAgeGreaterThan	…where x.age>?1
GreaterThanEqual	findByAgeGreaterThanEqual	…where x.age>=?1
After	findByStartDateAfter	…where x.startDate>?1
Before	findByStartDateBefore	…where x.startDate<?1
IsNull	findByAgeIsNull	…where x.age is null
IsNotNull,NotNull	findByAge(Is)NotNull	…where x.age not null
Like	findByFirstnameLike	…where x.firstname like ?1
NotLike	findByFirstnameNotLike	…findByFirstnameNotLike
StartingWith	findByFirstnameStartingWith	…where x.firstname like ?1(parameter bound with appended %)
EndingWith	findByFirstnameEndingWith	…where x.firstname like ?1
Containing	findByFirstnameContaining	…where x.firstname like ?1(parameter bound wrapped in %)
OrderBy	findByAgeOrderByLastnameDesc	…where x.age=?1 order by x.lastname desc
Not	findByLastnameNot	…where x.lastname<>?1
In	findByAgeIn(Collection<Age> ages)	…where x.age in ?1
NotIn	findByAgeNotIn(Collection<Age> ages)	… where x.age not in ?1
True	findByActiveTrue()	…where x.active=true
False	findByActiveFalse()	…where x.active=false
IgnoreCase	findByFirstnameIgnoreCase	…where UPPER(x.firstame)=UPPER(?1)

9. 自定义查询 Using @Query

既定的方法命名规则不一定满足所有的开发需求,因此 Spring Data JPA 也支持自定义 JPQL 或者原生 SQL。只需在声明的方法上面标注 @Query 注解,同时提供一个 JPQL 查询语句即可。如需对 username 进行模糊查询,则需在 UserRepository 中添加如下代码:

```
@Query("select u from User u where u.username like %?1%")
List<User> getUserByUsername(@Param("username") String username);
```

接着在 UserController 中暴露接口,代码如下:

```
@GetMapping("/ops")
public List<User> getOPs( String username){
    return userRepository.getUserByUsername(username);
}
```

15.3.3 CORS 跨域配置

在实际开发过程中我们经常会遇到跨域问题,跨域其实是浏览器对于 Ajax 请求的一种安全限制。目前比较常用的跨域解决方案有 3 种。

(1) Jsonp:最早的解决方案,利用 script 标签可以跨域的原理实现。
(2) nginx 反向代理:利用 nginx 反向代理把跨域转为不跨域,支持各种请求方式。
(3) CORS:规范化的跨域请求解决方案,安全可靠。

这里采用 CORS 的跨域方案。

CORS 全称"跨域资源共享"(Cross-Origin Resource Sharing)。CORS 需要浏览器和服务器同时支持,才可以实现跨域请求,目前绝大多数浏览器支持 CORS,IE 版本则不能低于 IE 10。CORS 的整个过程都由浏览器自动完成,前端无须做任何设置,跟平时发送 Ajax 请求并无差异。实现 CORS 的关键在于服务器(后台),只要服务器实现 CORS 接口,就可以实现跨域通信。

CORS 实现跨域也是非常简单的,事实上,Spring MVC 已经帮我们写好了 CORS 的跨域过滤器:CorsFilter,内部已经实现了判定逻辑,我们直接使用就可以了。

1. 配置 CORS

在 15.3.2 节项目的基础上进行前后端分离开发,在该项目下新建 GlobalCorsConfig 类用于跨域的全局配置,代码如下:

```
@Configuration
public class GlobalCorsConfig {
    @Bean
    public CorsFilter corsFilter() {
        //1. 添加 CORS 配置信息
        CorsConfiguration config = new CorsConfiguration();
```

```java
            //1) 允许接收的域,不要写 *,否则 cookie 就无法使用了
            config.addAllowedOrigin("http://localhost:8081");
            //2) 是否发送 Cookie 信息
            config.setAllowCredentials(true);
            //3) 允许的请求方式
            config.addAllowedMethod("OPTIONS");
            config.addAllowedMethod("HEAD");
            config.addAllowedMethod("GET");
            config.addAllowedMethod("PUT");
            config.addAllowedMethod("POST");
            config.addAllowedMethod("DELETE");
            config.addAllowedMethod("PATCH");
            //4)允许的头信息
            config.addAllowedHeader("*");
            //2.添加映射路径,我们拦截一切请求
            UrlBasedCorsConfigurationSource configSource = new UrlBasedCorsConfigurationSource();
            configSource.registerCorsConfiguration("/**", config);
            //3.返回新的 CorsFilter.
            return new CorsFilter(configSource);
    }
}
```

配置完成后启动该项目。

2．新建项目

新建一个 Spring Boot Web 项目,然后在 resources/static 目录下引入 jquery.js,再在 resources/static 目录下创建 index.html 文件,代码如下：

```html
<!DOCTYPE html>
<html lang="en">
<head>
    <meta charset="UTF-8">
    <title>Title</title>
    <script src="jquery-1.8.3.js"></script>
</head>
<body>
<input type="button" value="添加数据" onclick="addUser()"><br>
<input type="button" value="删除数据" onclick="deleteUser()">
<script>
    function addUser() {
        $.ajax({
            type:'post',
            /*访问 15.3.2 节项目中的添加接口*/
            url:'http://localhost:8080/users',
```

```
                    data:JSON.stringify({
                        "username":"admin",
                        "password":"123456",
                        "email":"635498720"
                    }),
                    dataType: "json",
                    contentType:"application/json;charset=UTF-8",
                    success:function (msg) {
                    }
                })
            }
            function deleteUser() {
                $.ajax({
                    type:'delete',
                    /*访问15.3.2节项目中的删除接口*/
                    url:'http://localhost:8080/users/5',
                    success:function (msg) {
                    }
                })
            }
</script>
</body>
</html>
```

在该文件中两个 Ajax 分别发送了跨域请求,请求 15.3.2 节项目中的添加和删除接口。然后将该项目的端口设置为 8081,代码如下:

```
server:
  port: 8081
```

3. 测试

启动项目,在浏览器中输入 http://localhost:8081/index.html,查看页面如图 15-10 所示。

分别单击图 15-10 所示的两个按钮,查看数据库中数据的变化。

图 15-10 index 页面

总的来说,CORS 的配置完全在后端设置,配置起来也比较容易,目前对于大部分浏览器兼容性也比较好。CORS 的优势也比较明显,可以实现任何类型的请求,相较于 Jsonp 跨域只能使用 GET 请求来说,也更加便于我们使用。

15.4　RESTful 风格

RESTful 是一种软件架构风格、设计风格,而不是标准,只是提供了一组设计原则和约束条件。它主要用于客户端和服务器交互类的软件。基于这个风格设计的软件可以更简洁,更有层次,更易于实现缓存等机制。在 15.3.2 节中展示了 RESTful 在 Spring Boot 中的应用开发。

我们知道以前网页是前端后端融合在一起的,例如 PHP、JSP 等。在之前的纯桌面时代问题不大,但是近年来在移动互联网的大潮下,随着 docker 等技术的兴起,微服务的概念也越来越被大家接受并应用于实践,各种类型的 Client 层出不穷,日益增多的 Web Service 逐渐统一于 RESTful 架构风格,RESTful 可以通过一套统一的接口为 Web、iOS 和 Android 提供服务,一般我们最常用的方式就是暴露 JSON 接口。

传统方式操作资源如下。

(1) 网站 http://localhost/item/queryUser.action?id=1 用于查询。

(2) 网站 http://localhost/item/saveUser.action 用于新增。

(3) 网站 http://localhost/item/updateUser.action 用于更新。

(4) 网站 http://localhost/item/deleteUser.action?id=1 用于删除。

RESTful 架构风格规定数据的元操作,即 CRUD(create、read、update 和 delete,即数据的增、删、查、改)操作,分别对应于 HTTP 方法:GET 用来获取资源,POST 用来新建资源(也可以用于更新资源),PUT 用来更新资源,DELETE 用来删除资源,这样就统一了数据操作的接口,仅通过 HTTP 方法,就可以完成对数据的所有增、删、查、改工作。

使用 RESTful 操作资源如下。

(1)【GET】/users　　用于查询用户信息列表。

(2)【GET】/users/1001　　用于查看某个用户信息。

(3)【POST】/users　　用于新建用户信息。

(4)【PUT】/users/1001　　用于更新用户信息(全部字段)。

(5)【PATCH】/users/1001　　用于更新用户信息(部分字段)。

(6)【DELETE】/users/1001　　用于删除用户信息。

RESTful 是一种旧技术新风格。

第 16 章 Spring Boot 安全框架

Spring Security 的前身是 Acegi Security，是 Spring 项目组中用来提供安全认证服务的框架，相对于 Shiro 它的功能更加强大。Spring Security 为基于 J2EE 企业应用软件提供了全面安全服务，而且非常方便与 Spring 项目无缝集成。特别是在 Spring Boot 项目中加入 Spring Security 更是十分简单，因为 Spring Boot 为 Spring Security 提供了自动化配置方案，可以零配置使用 Spring Security。本篇我们介绍 Spring Security，以及 Spring Security 在 Web 应用中的使用。

16.1 认识 Spring Security

Spring Security 致力于为 Java 应用提供核心功能认证（Authentication）和授权管理（Authorization）。

认证：主要为了解决我是谁的问题，通过提供证据证明你是你说的那个人。

授权：主要是为了解决我能干什么的问题。

它是一个强大的、高度自定义的认证和访问控制框架。其核心就是一组过滤器链，项目启动后将会自动配置。最核心的就是 Basic Authentication Filter 用来认证用户的身份，在 Spring Security 中一种过滤器处理一种认证方式，如图 16-1 所示。

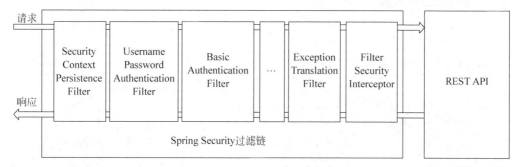

图 16-1 Spring Security 基本原理

16.1.1 入门项目

1. 创建项目

创建一个 Spring Boot 的 Web 项目,添加 Spring Security 的依赖,代码如下:

```xml
<dependency>
    <groupId>org.springframework.boot</groupId>
    <artifactId>spring-boot-starter-security</artifactId>
</dependency>
```

2. 控制器

新建 HelloController 类,添加 /hello 接口,代码如下:

```java
@RestController
public class HelloController {
    @RequestMapping("/hello")
    public String home() {
        return "Hello,spring security!";
    }
}
```

图 16-2 登录页面

3. 测试

启动项目,在浏览器访问 http://localhost:8080/hello 接口时会自动跳转到登录页面,这是 Spring Security 提供的,如图 16-2 所示。

说明 Spring Security 已经起作用了,它会把项目里的资源保护起来。Spring Security 默认的用户名是 user,Spring Security 启动的时候会生成默认密码,在启动日志中可以看到,如图 16-3 所示。

填入用户名和密码后,就可以成功访问接口了。

```
Using generated security password: 04329ef2-b1df-47cc-be3b-8c5d938ea98d
```

图 16-3 默认密码

4. 自定义用户名和密码

当然我们也可以自己设置用户名和密码,在 application.yml 中添加如下配置:

```yaml
spring:
  security:
```

```yaml
user:
  name: admin          #用户名
  password: 123456     #密码
```

重启项目,访问被保护的/hello 接口。自动跳转到了 Spring Security 默认的登录页面,输入用户名 admin 和密码 123456 后成功跳转到了/hello。

16.1.2 角色访问控制

通常情况下,我们需要实现"特定资源只能由特定角色访问"的功能。假设我们的系统有以下两个角色。

(1) ADMIN:可以访问所有资源。
(2) USER:只能访问特定资源。

现在给系统增加"/user/**"接口代表用户信息方面的资源(USER 可以访问);增加"/admin/**"接口代表管理员方面的资源(USER 不能访问),代码如下:

```java
/*用户信息方面的资源*/
@RestController
public class UserController {
    @RequestMapping("/user/hello")
    public String hello() {
        return "user,Hello!";
    }
}
/*管理员方面的资源*/
@RestController
public class AdminController {
    @RequestMapping("/admin/hello")
    public String hello() {
        return "admin,Hello!";
    }
}
```

在实际开发中,用户和角色是保存在数据库中的,本例为了方便演示,我们来创建两个存放于内存的用户和角色。可以自定义类并集成 WebSecurityConfigurerAdapter 进而实现 Spring Security 的更多配置,代码如下代码:

```java
@Configuration
public class SecurityConfig extends WebSecurityConfigurerAdapter {
    /*不对密码进行加密*/
    @Bean
    PasswordEncoder passwordEncoder(){
```

```java
            return NoOpPasswordEncoder.getInstance();
    }
    @Override
    protected void configure(AuthenticationManagerBuilder auth) throws Exception {
        auth.inMemoryAuthentication()
                /*管理员用户,具备 ADMIN 和 USER 角色*/
                .withUser("admin").password("admin").roles("ADMIN", "USER")
                .and()
                /*普通用户*/
                .withUser("beixi").password("beixi").roles("USER");
    }
    @Override
    protected void configure(HttpSecurity http) throws Exception {
        http
                .authorizeRequests()
                /*普通用户访问的 URL*/
                .antMatchers("/user/**").hasRole("USER")
                /*管理员用户访问的 URL*/
                .antMatchers("/admin/**").hasRole("ADMIN")
                .anyRequest().authenticated()          //其他路径都必须认证
                .and()
                .formLogin()
                .loginProcessingUrl("/login")
                .permitAll()       //访问/login 接口不需要进行身份认证了,防止重定向死循环
                .and()
                .csrf().disable();                     //关闭 csrf
    }
}
```

根据上面的配置,我们知道使用"beixi"用户具有访问"/user/**"接口的权限,使用 ADMIN 登录,可以访问所有接口。

16.2 基于数据库的认证

在 16.1 节中,用户登录系统的用户名、密码定义在内存中,在实际开发中,用户的基本信息及角色都是通过查询数据库进行认证和授权。

16.2.1 Spring Security 基于数据库认证

1. 创建表

在 MySQL 数据库中创建一张用户表,id 为主键自增,并添加两个用户,代码如下:

```sql
DROP TABLE IF EXISTS 'user';
CREATE TABLE 'user' (
```

```
    'id' INT NOT NULL AUTO_INCREMENT COMMENT '主键',
    'username' VARCHAR(32) DEFAULT '' COMMENT '用户姓名',
    'password' VARCHAR(100) DEFAULT '' COMMENT '密码',
    'role' VARCHAR(32) COMMENT '角色',,
    PRIMARY KEY ('id')
);
##添加用户信息
INSERT INTO 'user' VALUES (1, 'user', '123','user');
INSERT INTO 'user' VALUES (2, 'admin', '123','admin');
```

2．创建项目

创建一个 Spring Boot 模块项目，选择相关依赖，如图 16-4 所示。

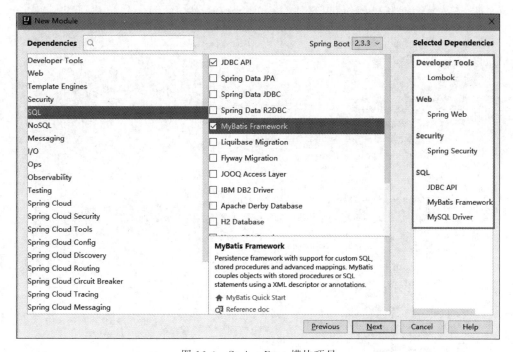

图 16-4　Spring Boot 模块项目

Lombok 的主要作用是通过注解消除实际开发中的样板式代码，如：getter、setter 方法，重写 toString、equals 方法等，这些代码没有什么技术含量，但是常常要写，因此可以用 @Data、@Setter、@Getter 等注解来替换。

3．配置文件

在 application.yml 中设置 MySQL 连接配置和 MyBatis 配置，代码如下：

```
spring:
  datasource:
```

```yaml
    url: jdbc:mysql://localhost:3306/jpa?serverTimezone=GMT%2B8&characterEncoding=utf-8
    username: root
    password: root
    driver-class-name: com.mysql.jdbc.Driver
logging:
  level:
    com.example.bdatabaserole.mapper: debug     # 打印SQL语句
```

4. 创建实体类

创建与用户表相对应的实体类并实现 UserDetails,代码如下:

```java
@Data      //Lombok 注解省略 get/set 等方法
public class UserInfo implements Serializable,UserDetails {
    private int id;
    private String username;
    private String password;
    private String role; //角色

    public UserInfo(String username, String password, String role) {
        this.username = username;
        this.password = password;
        this.role = role;
    }
    /**
     * 指示用户的账户是否已过期。无法验证过期的账户。
     * 如果用户的账户有效(即未过期),则返回true,如果不再有效就返回false
     */
    @Override
    public boolean isAccountNonExpired() {
        return true;
    }
    /**
     * 指示用户是锁定还是解锁。无法对锁定的用户进行身份验证。
     * 如果用户未被锁定,则返回true,否则返回false
     */
    @Override
    public boolean isAccountNonLocked() {
        return true;
    }
    /**
     * 指示用户的凭证(密码)是否已过期。过期的凭证阻止身份验证
     * 如果用户的凭证有效(即未过期),则返回true
     * 如果不再有效(即过期),则返回false
     */
```

```java
    @Override
    public boolean isCredentialsNonExpired() {
        return true;
    }
    /**
     * 指示用户是启用还是禁用。无法对禁用的用户进行身份验证
     * 如果启用了用户,则返回 true,否则返回 false
     */
    @Override
    public boolean isEnabled() {
        return true;
    }
    /**
     * 得到用户的权限,如果权限表和用户表是分开的,我们需要再重新定义一个实体类实现
     * UserDetails 并且继承于 User 类
     * 交给 security 的权限,放在 UserDetailService 进行处理
     */
    @Override
    public Collection<? extends GrantedAuthority> getAuthorities() {
        Collection<GrantedAuthority> authorities = new ArrayList<>();
        /* 角色必须以 ROLE_ 开头,如果数据库没有,则需要加上 */
        authorities.add(new SimpleGrantedAuthority("ROLE_" + this.role));
        return authorities;
    }
}
```

代码解释:

UserDetails 接口是提供用户信息的核心接口。该接口仅仅实现存储用户的信息。后续会将该接口提供的用户信息封装到认证对象 Authentication 中去。UserDetails 默认提供了 7 个方法,如表 16-1 所示。

表 16-1 UserDetails 提供的 7 个方法

方法名	作用
getAuthorities()	用户的角色集,默认需要添加 ROLE_ 前缀
getPassword()	获取当前用户对象密码
getUsername()	获取当前用户对象用户名
isAccountNonExpired()	账户是否过期
isAccountNonLocked()	账户是否锁定
isCredentialsNonExpired()	凭证是否过期
isEnabled()	用户是否可用

5. 创建 Mapper 接口和 service 层

创建 UserMapper 接口,对于简单的 SQL 语句可以使用注解式来替换 XML 式,接着创

建 UserInfoService 业务层类,代码如下:

```java
@Mapper
@Repository
public interface UserMapper {
    @Select("select * from user where username = #{username}")
    UserInfo getUserByUsername(String username);
}

@Service
public class UserInfoService {
    @Autowired
    private UserMapper userMapper;
    public UserInfo getUserInfo(String username){
        return userMapper.getUserByUsername(username);
    }
}
```

6. 创建 controller

接下来创建 UserController,代码如下:

```java
@RestController
public class UserController {
    @Autowired
    private UserInfoService userInfoService;
    @GetMapping("/getUser")
    public UserInfo getUser(@RequestParam String username){
        return userInfoService.getUserInfo(username);
    }
}
```

7. 身份认证

要从数据库读取用户信息进行身份认证,需要新建类实现 UserDetailService 接口并重写 loadUserByUsername 方法,该方法的参数是登录时的用户名,通过该用户名去数据库查找用户,如果不存在就抛出用户不存在的异常,如果查到了用户,就会将用户及角色信息返回。loadUserByUsername 方法将在用户登录时自动调用,代码如下:

```java
@Component
public class CustomUserDetailsService implements UserDetailsService {
    @Autowired
    private UserInfoService userInfoService;
    /**
     * 需新建配置类注册一个指定的加密方式 Bean,或在下一步 Security 配置类中注册指定
     */
```

```java
    @Autowired
    private PasswordEncoder passwordEncoder;

    @Override
    public UserDetails loadUserByUsername(String username) throws UsernameNotFoundException
{
        //通过用户名从数据库获取用户信息
        UserInfo userInfo = userInfoService.getUserInfo(username);
        if (userInfo == null) {
            throw new UsernameNotFoundException("用户不存在");
        }
        //得到用户角色
        String role = userInfo.getRole();
        //角色集合
        List<GrantedAuthority> authorities = new ArrayList<>();
        //角色必须以'ROLE_'开头,如果数据库中没有,则在这里添加
        authorities.add(new SimpleGrantedAuthority("ROLE_" + role));
        return new User(
                userInfo.getUsername(),
                //因为数据库是明文,所以这里需要加密密码
                passwordEncoder.encode(userInfo.getPassword()),
                authorities
        );
    }
}
```

8. Spring Security 的配置

创建 WebSecurityConfig 继承 WebSecurityConfigurerAdapter,并重写 configure(auth) 方法:

```java
@EnableWebSecurity       //Spring Security用于启用Web安全的注解
public class WebSecurityConfig extends WebSecurityConfigurerAdapter {
    @Autowired
    private CustomUserDetailsService userDatailService;
    /**
     * 指定加密方式
     */
    @Bean
    public PasswordEncoder passwordEncoder(){
        // 使用BCrypt加密密码
        return new BCryptPasswordEncoder();
    }
    @Override
    protected void configure(AuthenticationManagerBuilder auth) throws Exception {
```

```
        auth
                //从数据库读取的用户进行身份认证
                .userDetailsService(userDatailService)
                .passwordEncoder(passwordEncoder());
    }
}
```

9．测试

配置完成后，启动项目，在登录页面输入数据库中的用户名、密码，成功后就可以访问接口了，如图16-5所示。

```
{
    id: 1,
    username: "user",
    password: "123",
    role: "user",
    enabled: true,
  - authorities: [
        - {
              authority: "ROLE_user"
          }
    ],
    accountNonExpired: true,
    accountNonLocked: true,
    credentialsNonExpired: true
```

图16-5　成功访问接口

16.2.2　角色访问控制

上面设置完成后，可以使用数据库中的用户名和密码登录，并获得用户的角色。接下来要通过用户的角色，限制用户的请求访问。

1．开启访问权限

在16.1节中，角色访问控制是基于URL配置的，我们也可以通过注解来灵活地配置方法安全，在WebSecurityConfig添加@EnableGlobalMethodSecurity注解开启方法的访问权限，代码如下：

```
@EnableWebSecurity //Spring Security 用于启用 Web 安全的注解
@EnableGlobalMethodSecurity(prePostEnabled = true) //开启方法级安全验证
public class WebSecurityConfig extends WebSecurityConfigurerAdapter {
    //...
}
```

代码解释：

prePostEnabled＝true 会解锁@PreAuthorize 和@PostAuthorize 两个注解，@PreAuthorize

注解会在方法执行前进行验证，而@PostAuthorize 注解在方法执行后进行验证。

2．在控制层添加访问接口

在 UserController 类中增加方法的访问权限：

```
@RestController
public class UserController {
    @Autowired
    private UserInfoService userInfoService;
    @GetMapping("/getUser")
    public UserInfo getUser(@RequestParam String username){
        return userInfoService.getUserInfo(username);
    }
    @PreAuthorize("hasAnyRole('user')")           //只有 user 角色才能访问该方法
    @GetMapping("/user")
    public String user(){
        return "hello,user";
    }
    @PreAuthorize("hasAnyRole('admin')")          //只有 admin 角色才能访问该方法
    @GetMapping("/admin")
    public String admin(){
        return "hello,admin";
    }
}
```

代码解释：

@PreAuthorize("hasAnyRole('user')")注解表示访问该方法需要 user 角色。

3．测试

重新启动程序，使用角色为 user 的用户登录，可以访问 localhost:8080/user，不能访问 localhost:8080/admin。如果使用角色为 admin 的用户登录，则可以访问所有。

16.2.3　密码加密保存

上文中的用户密码都是手动在数据库中添加的，所以在数据库中以明文显示，在实际开发中，用户密码都需要加密保存。下面模拟注册用户，并加密保存密码。

1．修改 Mapper 接口

在 UserMapper 接口中添加插入用户，代码如下：

```
@Mapper
@Repository
public interface UserMapper {
//...
    @Insert("insert into user(username, password, role) value(#{username}, #{password}, #{role})")
    int insertUserInfo(UserInfo userInfo);
}
```

2. 修改 service 类

在 UserInfoService 类中添加插入方法,并且密码要加密保护,代码如下:

```java
@Service
public class UserInfoService {
    //...
    @Autowired
    private PasswordEncoder passwordEncoder;
    //...
    public int insertUser(UserInfo userInfo){
        /* 加密密码 */
        userInfo.setPassword(passwordEncoder.encode(userInfo.getPassword()));
        return userMapper.insertUserInfo(userInfo);
    }
}
```

3. 修改 controller

在 UserController 类中添加插入用户接口,代码如下:

```java
@RestController
public class UserController{
    //...
    @PostMapping("/addUser")
    public int addUser(@RequestBody UserInfo userInfo){
        return userInfoService.insertUser(userInfo);
    }
}
```

4. 测试

配置完后,启动服务,使用 Postman 发送 POST 请求来添加用户,如图 16-6 所示。

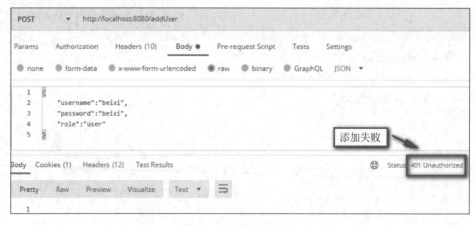

图 16-6　添加用户失败

单击 Send 按钮后,由图 16-6 可以看出添加失败,响应的状态码显示 401 Unauthorized,说明无权限,需要登录,但注册用户是不需要登录的,所以需要给注册用户释放权限。

修改 WebSecurityConfig 配置类,重写 configure(HttpSecurity http)方法,配置允许注册用户的请求访问:

```java
@EnableWebSecurity
@EnableGlobalMethodSecurity(prePostEnabled = true)
public class WebSecurityConfig extends WebSecurityConfigurerAdapter {
    //...
    @Override
    protected void configure(HttpSecurity http) throws Exception {
        http
            .authorizeRequests()
            //允许 POST 请求/addUser,而无须认证
            .antMatchers(HttpMethod.POST, "/addUser").permitAll()
            .anyRequest().authenticated()          //所有请求都需要验证
            .and()
            .formLogin()                           //使用默认的登录页面
            .and()
            .csrf().disable();
//POST 请求要关闭 csrf 验证,不然访问报错;实际开发中开启,需要前端配合传递其他参数
    }}
```

5. 重启项目

再次访问/addUser 接口,可以看到 Postman 发送请求成功了,如图 16-7 所示。

图 16-7　添加用户成功

查看数据库数据,添加的用户密码已加密,如图 16-8 所示。

6. 使用加密密码登录

使用加密密码登录,需要修改 CustomUserDetailsService 类,之前从数据库获取明文密

id	username	password	role
1	user	123	user
2	admin	123	admin
3	beixi	$2a$10$8Imm88K/vRajnr4KpTSnFuMoTip.8N9NW1wLvKDiShAQXYxfFktBS	user

图 16-8　user 表数据

码后需要加密，现在数据库里面的密码已经加密了，就不用再加密了，代码如下：

```java
@Component
public class MyUserDatailService implements UserDetailsService {
//...
    @Override
    public UserDetails loadUserByUsername(String username) throws UsernameNotFoundException
{
//...
        return new User(
                userInfo.getUsername(),
                //数据库密码已加密,不用再加密
                userInfo.getPassword(),
                authorities
        );
    }
}
```

在浏览器访问 localhost:8080/user，输入 beixi/beixi 登录即可。

16.2.4　用户角色多对多关系

认证数据是多个用户对一个角色的，在实际项目中，很多时候都是多对多的情况，我们在该项目的基础上进行开发讲解，本节将向读者介绍用户角色多对多关系的数据认证和授权。

1．创建表结构

一共三张表，分别是用户表、角色表及用户角色关联表，代码如下：

```sql
DROP TABLE IF EXISTS 'role';
CREATE TABLE 'role' (
    'id' INT(11) NOT NULL AUTO_INCREMENT,
    'name' VARCHAR(32) DEFAULT NULL,
    'nameZh' VARCHAR(32) DEFAULT NULL,
    PRIMARY KEY ('id')
) ENGINE = INNODB AUTO_INCREMENT = 4 DEFAULT CHARSET = utf8;

INSERT INTO 'role' VALUES ('1', 'ROLE_dba', '数据库管理员');
INSERT INTO 'role' VALUES ('2', 'ROLE_admin', '系统管理员');
INSERT INTO 'role' VALUES ('3', 'ROLE_user', '用户');
```

```sql
DROP TABLE IF EXISTS 'user';
CREATE TABLE 'user' (
    'id' INT(11) NOT NULL AUTO_INCREMENT,
    'username' VARCHAR(32) DEFAULT NULL,
    'password' VARCHAR(255) DEFAULT NULL,
    PRIMARY KEY ('id')
) ENGINE = INNODB AUTO_INCREMENT = 4 DEFAULT CHARSET = utf8;

INSERT INTO 'user' VALUES ('1', 'root', '$2a$10$em/RtpOT.I1heN/9dRof0uRuUwB5k8uzN18l73Wpah6KcMydr5tCK');
INSERT INTO 'user' VALUES ('2', 'admin', '$2a$10$em/RtpOT.I1heN/9dRof0uRuUwB5k8uzN18l73Wpah6KcMydr5tCK');
INSERT INTO 'user' VALUES ('3', 'beixi', '$2a$10$em/RtpOT.I1heN/9dRof0uRuUwB5k8uzN18l73Wpah6KcMydr5tCK');

DROP TABLE IF EXISTS 'user_role';
CREATE TABLE 'user_role' (
    'id' INT(11) NOT NULL AUTO_INCREMENT,
    'uid' INT(11) DEFAULT NULL,
    'rid' INT(11) DEFAULT NULL,
    PRIMARY KEY ('id')
) ENGINE = INNODB AUTO_INCREMENT = 5 DEFAULT CHARSET = utf8;
INSERT INTO 'user_role' VALUES ('1', '1', '1');
INSERT INTO 'user_role' VALUES ('2', '1', '2');
INSERT INTO 'user_role' VALUES ('3', '2', '2');
INSERT INTO 'user_role' VALUES ('4', '3', '3');
SET FOREIGN_KEY_CHECKS = 1;
```

其中密码是 123，Spring Security 加密算法后的密文，可以让同样的密码每次生成的密文都不一样，从而不会重复，这里是我复制后粘贴的所以密文一样。

注意：角色名前需要加 ROLE_ 前缀。

2. 创建实体类

根据表结构修改 UserInfo 实体类，创建 Role 类，代码如下：

```java
@Data
public class UserInfo implements Serializable,UserDetails {
//...
    private List<Role> roleList;
    @Override
    public Collection<? extends GrantedAuthority> getAuthorities() {
        Collection<GrantedAuthority> authorities = new ArrayList<>();
        for (Role role : roleList) {
            //数据库 role 表字段中是以 ROLE_ 开头的,所以此处不必再加 ROLE_
            authorities.add(new SimpleGrantedAuthority(role.getName()));
```

```
        }
        return authorities;
    }
}
@Data
public class Role {
    private int id;
    private String name;
    private String nameZh;
}
```

3. 创建 UserMapper 和 UserMapper.xml

修改 UserMapper 接口中的抽象方法,代码如下:

```
@Mapper
@Repository
public interface UserMapper {
    UserInfo getUserByUsername(String username);
    List<Role> getRolesById(int id);
//...
}
```

接着在 classpath:mappers 下创建 UserMapper.xml 映射文件,当 SQL 语句比较复杂时,建议使用 XML 式,代码如下:

```
<?xml version="1.0" encoding="UTF-8"?>
<!DOCTYPE mapper
        PUBLIC "-//mybatis.org//DTD Mapper 3.0//EN"
"http://mybatis.org/dtd/mybatis-3-mapper.dtd">
<!--命名空间必须和 UserMapper 全类名相同-->
<mapper namespace="com.beixi.mapper.UserMapper">
    <select id="getUserByUsername" resultType="com.beixi.entity.UserInfo">
        select * from user where username = #{username};
    </select>
    <select id="getRolesById" resultType="com.beixi.entity.Role">
        select * from role where id in(select rid from user_role where uid = #{uid});
    </select>
</mapper>
```

并在 application.yml 中进行配置,代码如下:

```
#...
mybatis:
```

```yaml
  mapper-locations: classpath:mappers/*.xml
  type-aliases-package: com.beixi.mapper
```

4. 修改 CustomUserDetailsService 类

修改 CustomUserDetailsService 类中 loadUserByUsername 方法，代码如下：

```java
@Component
public class CustomUserDetailsService implements UserDetailsService {
    //...
    @Autowired
    private UserMapper userMapper;
    @Override
    public UserDetails loadUserByUsername(String username) throws UsernameNotFoundException
    {
        //通过用户名从数据库获取用户信息
        UserInfo userInfo = userInfoService.getUserInfo(username);
        if (userInfo == null) {
            throw new UsernameNotFoundException("用户不存在");
        }
        userInfo.setRoleList(userMapper.getRolesById(userInfo.getId()));
        return userInfo;
    }
}
```

5. controller 层

在 UserController 类中添加如下方法：

```java
@RestController
public class UserController {
    //...
    @PreAuthorize("hasAnyRole('user')")          //只有 user 角色才能访问该方法
    @GetMapping("/user")
    public String user(){
        return "hello,user";
    }
    @PreAuthorize("hasAnyRole('dba','admin')")   // dba 或 admin 角色可以访问该方法
    @GetMapping("/db")
    public String dba(){
        return "hello,dba,admin";
    }
    @PreAuthorize("hasAnyRole('admin')")         //只有 admin 角色才能访问该方法
    @GetMapping("/admin")
    public String admin(){
        return "hello,admin";
    }
}
```

配置完成后,启动项目对 controller 中的接口进行测试,方式跟上文中的一样,此处不再赘述。

16.2.5 角色继承

角色继承实际上是一个很常见的需求,因为大部分公司可能采用金字塔形的治理方式,上司可能具备下属的部分甚至所有权限,这一现实场景,反映到我们的代码中,就是角色继承了。

Spring Security 中为开发者提供了相关的角色继承解决方案,只需开发者在配置类中提供一个 RoleHierarchy,代码如下:

```
@Bean
RoleHierarchy roleHierarchy() {
    RoleHierarchyImpl roleHierarchy = new RoleHierarchyImpl();
    String hierarchy = "ROLE_dba > ROLE_admin \n ROLE_admin > ROLE_user";
    roleHierarchy.setHierarchy(hierarchy);
    return roleHierarchy;
}
```

这里提供了一个 RoleHierarchy 接口的实例,使用字符串来描述了角色之间的继承关系,ROLE_dba 具备 ROLE_admin 的所有权限,而 ROLE_admin 则具备 ROLE_user 的所有权限。提供了这个 Bean 之后,以后所有具备 ROLE_user 角色才能访问的资源,ROLE_dba 和 ROLE_admin 也都能访问,具备 ROLE_amdin 角色才能访问的资源,ROLE_dba 也能访问。

第 17 章 项目构建与部署

Spring Boot 项目可以内嵌 Servlet 容器，因此部署极为方便，可直接打包成可执行 Jar 包部署在有 Java 运行环境的服务器上，也可以打成 War 包并部署到外部 Tomcat 服务器上等。

17.1 Jar 部署

Spring Boot 打包成 Jar 包一般使用 spring-boot-maven-plugin 这个插件，该插件在创建 Spring Boot Web 项目时自动会在 pom.xml 文件中生成，代码如下：

```xml
<plugin>
    <groupId>org.springframework.boot</groupId>
    <artifactId>spring-boot-maven-plugin</artifactId>
</plugin>
```

当配置了该插件后就可以创建一个可执行的 Jar 文件。这在很大程度上简化了应用的部署，只需安装了 JRE 就可以运行。不过前提是应用程序的 parent 为 spring-boot-starter-parent。

配置完成后在 Maven Project 中双击执行 clean，当 clean 执行完毕后，再执行 install，如图 17-1 所示。执行完毕后，会发现在项目根目录的 target 目录下有刚刚打好的 Jar 包。

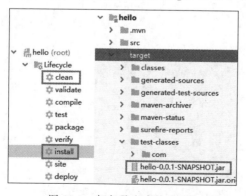

图 17-1　打包及生成的 Jar 包

接着直接进入 target 目录下执行如下命令即可启动项目：

```
java -jar hello-0.0.1-SNAPSHOT.jar
```

或者把这个 Jar 文件放在任意盘符中，按住 Shift＋鼠标右键进入 PowerShell 输入命令启动项目，如图 17-2 所示。

图 17-2

17.2 War 部署

Spring Boot 默认打包成 Jar 包，使用 Spring Boot 构造 Web 应用，默认使用内置的 Tomcat。但考虑到项目需要集群部署或者进行优化，就需要打包成 War 包部署到外部的 Tomcat 服务器中。

1. 修改打包形式

修改 pom.xml 文件将默认的 Jar 方式改为 War 方式，代码如下：

```xml
<packaging>war</packaging>
```

2. 排除内置的 Tomcat 容器

在 pom.xml 文件中移除 Tomcat，代码如下：

```xml
<dependency>
    <groupId>org.springframework.boot</groupId>
    <artifactId>spring-boot-starter-web</artifactId>
    <exclusions>
        <exclusion>
            <groupId>org.springframework.boot</groupId>
            <artifactId>spring-boot-starter-tomcat</artifactId>
        </exclusion>
    </exclusions>
</dependency>
```

3. 添加 servlet-api 依赖

在 pom.xml 文件中添加 servlet-api 依赖，代码如下：

```xml
<dependency>
    <groupId>javax.servlet</groupId>
    <artifactId>javax.servlet-api</artifactId>
    <version>3.1.0</version>
    <scope>provided</scope>
</dependency>
```

4. 修改启动类

启动类继承 SpringBootServletInitializer 并重写 configure 方法，代码如下：

```java
/* 修改启动类,继承 SpringBootServletInitializer 并重写 configure 方法 */
@Override
protected SpringApplicationBuilder configure(SpringApplicationBuilder builder) {
    //注意这里要指向原先用 main 方法执行的 Application 启动类
    return builder.sources(Application.class);
}
```

5. 打包部署

打包 War 包的方式和打包 Jar 包的方式一样都是在 Maven Project 中双击执行 clean，当 clean 执行完毕后，再执行 install。然后把 target 目录下的 War 包放到 Tomcat 的 webapps 目录下，启动 Tomcat，即可自动解压部署。最后在浏览器中就可以访问该打包项目。

注意：使用外部 Tomcat 部署访问时，application.properties（或者 application.yml）中的配置将失效，请使用外置 Tomcat 的端口和 webapps 目录下项目名进行访问。

第 18 章 部门管理系统

从经典的 JSP + Servlet + JavaBean 的 MVC 时代,到 SSM(Spring + Spring MVC + MyBatis)和 SSH(Spring + Struts + Hibernate)的 Java 框架时代,再到前端框架(KnockoutJS、AngularJS、Vue.js、ReactJS)为主的 MVVM 时代,然后是 Node.js 引领的全栈时代,技术和架构一直都在进步。创新之路不会止步,无论是前后端分离模式还是其他模式,都是为了更方便地解决需求问题,但它们都只是一个"中转站"。前端项目与后端项目是两个项目,放在两个不同的服务器,需要独立部署,两个不同的工程,两个不同的代码库,不同的开发人员。前端只需关注页面的样式与动态数据的解析及渲染,而后端则专注于具体业务逻辑。本章将通过一个前后端分离项目带大家掌握目前流行的 Vue.js + Spring Boot 前后端开发环境及项目的开发。

18.1 技术分析

前后端分离的核心思想是前端 HTML 页面通过 Ajax 调用后端的 RESTful API 并使用 JSON 数据进行交互。本项目采用前后端开发,后端使用 Spring Boot + JPA,前端使用 Vue.js + ElementUI 来构建 SPA。前后端分离并非仅仅只是前后端开发的分工,而是在开发期进行代码存放分离、开发职责分离,前后端能够独立进行开发测试。在运行期进行应用部署分离,前后端之间通过 HTTP 请求进行通信。前后端分离的开发模式与传统模式相比,能为我们提升开发效率、增强代码可维护性,让我们有规划地打造一个前后端并重的精益开发团队,更好地应对越来越复杂多变的 Web 应用开发需求。

18.2 项目构建

18.2.1 前端项目搭建

前端项目是基于 Node.js 环境基础上的,使用 Vue-cli 脚手架快速创建 Vue.js 项目,如创建一个名为 myProject 的前端项目,在 7.1 节已详细介绍,大家可以参考,这里不再赘述,在 cmd 中执行如图 18-1 所示命令即可。

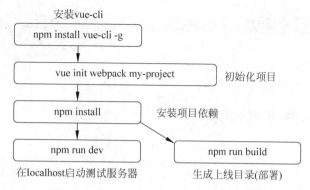

图 18-1　vue-cli 脚手架创建 Vue 项目命令

安装并启动成功后在浏览器输入 http://localhost:8080，显示页面如图 18-2 所示。

图 18-2　前端项目启动成功页面

18.2.2　后端项目搭建

使用 Spring Initializer 快速创建 Spring Boot 模板项目，如项目名为 myProject，如图 18-3 所示。

在实际开发中，项目所需的依赖不止这些，可以根据需求在 Dependencies 选择自己所需的依赖或者手动添加。

18.2.3　数据库设计

采用轻量的关系型数据库 MySQL，本项目主要以带领大家体验目前流行的前后端开发环境及项目的开发为主，所以表结构也比较简单，如图 18-4 所示。

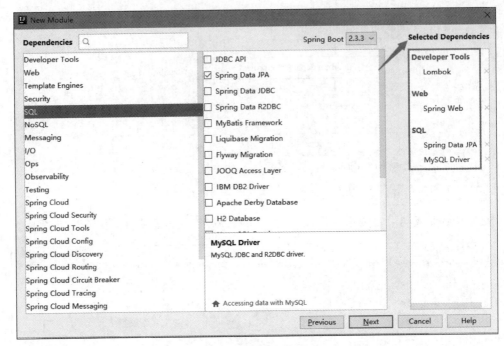

图 18-3　Spring Boot 模板项目

图 18-4　dept 表结构

至此，准备工作就基本搭建完毕了。

18.3　查询数据

18.3.1　后端实现

1. 配置文件

在 application.yml 文件中配置数据库连接、JPA 及端口等信息，代码如下：

```yaml
spring:
  datasource:
    #key: 后必须加空格,否则执行时报错
    url: jdbc:mysql://localhost:3306/test4?serverTimezone=CTT&useUnicode=true&characterEncoding=utf-8&allowMultiQueries=true
    username: root
    password: root
    driver-class-name: com.mysql.cj.jdbc.Driver
  jpa:
    show-sql: true
    properties:
      hibernate:
        formate_sql: true
server:
  port: 8081          #需要修改端口,否则和前端项目端口冲突
```

2. 实体类

配置完成后建立和表结构相对应的实体类 Dept,代码如下:

```java
@Entity
@Data
public class Dept {
    @Id
    @GeneratedValue(strategy = GenerationType.IDENTITY)
    private Integer deptno;
    private String dname;
    private String loc;
}
```

3. Dao 层

创建 DeptRepository 接口,并继承 JpaRepository 类,该类中封装了基本的增、删、改、查、分页、排序等方法,可参考 15.3.2 节,代码如下:

```java
public interface DeptRepository extends JpaRepository<Dept,Integer> {
}
```

4. controller 层

这里为了简化操作省略 service 层,在 DeptController 类中创建查询方法,代码如下:

```java
@RestController
@RequestMapping("/dept")
public class DeptController {
    @Autowired
```

```
    private DeptRepository deptRepository;
    @GetMapping("/findAll")
    public List<Dept> findAll(){
        return deptRepository.findAll();
    }
}
```

5. 测试

启动项目,使用 Postman 工具进行测试,如图 18-5 所示。

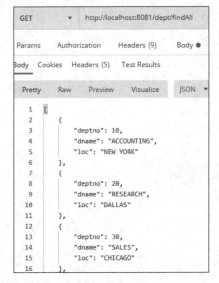

图 18-5　查询接口

12min

18.3.2　前端实现

1. 创建 Dept 页面

首先在 components 目录下创建部门静态数据模板 Dept.vue 页面,代码如下:

```
<template>
  <div>
    <table>
      <tr>
        <td>编号</td>
        <td>部门名称</td>
        <td>地址</td>
      </tr>
      <!-- 遍历数据 -->
```

```
        <tr v-for="item in depts">
            <td>{{item.deptno}}</td>
            <td>{{item.dname}}</td>
            <td>{{item.loc}}</td>
        </tr>
    </table>
  </div>
</template>
<script>
    export default {
        name: "Dept",
        data:function(){
          return {
              depts:[
                  {
                      deptno:1,
                      dname:'研发部',
                      loc:'北京'
                  },{
                      deptno:2,
                      dname:'人事部',
                      loc:'北京'
                  },{
                      deptno:3,
                      dname:'行政部',
                      loc:'北京'
                  }
              ]
          }
        }
    }
</script>
```

2. 修改路由

修改 router/index.js 文件的路由配置信息，代码如下：

```
import Vue from 'vue'
import Router from 'vue-router'
import HelloWorld from '@/components/HelloWorld'
import Dept from '../components/Dept.vue'
Vue.use(Router)
export default new Router({
  routes: [
    {
```

```
        path: '/', /* 首次直接访问部门页面 */
        name: 'Dept',
        component: Dept
      }
    ]
})
```

另外,由于 main.js 是入口 JS,在 main.js 中导入了 App 组件,App 组件中默认有 Vue.js 的 Logo,将 Logo 删除,只保留路由占位符即可,代码如下:

```
<template>
  <div id="app">
    <router-view/>
  </div>
</template>
<!--- 部分代码省略 -->
```

3. 测试

执行 npm run dev 命令,启动前端项目,启动成功后在浏览器输入 http://localhost:8080/,即可访问静态数据,如图 18-6 所示。

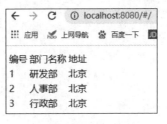

图 18-6 部门静态数据

4. 引入 Axios

首先安装 Axios 的依赖,命令如下:

```
npm install --save axios
```

依赖添加成功后,在 main.js 中引用并赋值 Axios,代码如下:

```
import axios from 'axios';          //加这句引用
Vue.prototype.$http = axios;        //加这句赋值
```

接着在 Dept 页面使用 Axios 请求后端的实时数据,代码如下:

```
<!-- 部分代码省略 -->
```

```
<script>
    export default {
        name: "Dept",
        data:function(){
          return {
              depts:[]
          }
        },
        created:function(){
         this.$http.get('http://localhost:8081/dept/findAll').then(resp=>{
             this.depts = resp.data; /*后端数据赋值给 depts 数组*/
         })
        }
    }
</script>
```

在前后端项目均启动的情况下,访问 http://localhost:8080/,此时会发现请求不到数据,控制台输出如图 18-7 所示的错误。

```
Access to XMLHttpRequest at 'http://localhost:8081/dept/findAl  :8080/#/:1
l' from origin 'http://localhost:8080' has been blocked by CORS policy: No
'Access-Control-Allow-Origin' header is present on the requested resource.
▶ GET http://localhost:8081/dept/findAll net::ERR_FAILED    xhr.js?ec6c:184
▶ Uncaught (in promise) Error: Network Error         createError.js?16d0:16
       at createError (createError.js?16d0:16)
       at XMLHttpRequest.handleError (xhr.js?ec6c:91)
```

图 18-7　错误信息

这个错误表明权限不足,因为前端项目和后端项目在不同的端口下启动,所以需要配置跨域请求。跨域处理本书讲了两种方法:一种是在前端使用代理的方式,可参考 8.3.3 节;另一种是后端使用 CORS 跨域配置,可参考 15.3.3 节。

这里采用后端 CORS 跨域配置,在后端项目中添加 CrosConfig 类,代码如下:

```
@Configuration
public class CrosConfig implements WebMvcConfigurer {
    @Override
    public void addCorsMappings(CorsRegistry registry) {
        registry.addMapping("/**")
                .allowedOrigins("*")
                .allowedMethods("GET","HEAD","POST","PUT","DELETE","OPTIONS")
                .allowCredentials(true)
                .maxAge(3600)
                .allowedHeaders("*");
    }
}
```

这样就可以成功访问后端的实时数据了。

18.4 加载菜单

18.4.1 引入ElementUI

数据成功访问后，引入ElementUI组件对数据进行渲染，首先引入其依赖：

```
npm install element-ui -S
```

依赖添加成功后，接着在main.js中引入ElementUI，代码如下：

```
import ElementUI from 'element-ui'
import 'element-ui/lib/theme-chalk/index.css'
Vue.use(ElementUI)
```

引入之后，在项目中就可以直接使用其相关组件了，官网地址为https://element.eleme.io/#/zh-CN/component/installation。

18.4.2 菜单

1. 菜单模板

菜单是用户成功访问后首页显示的项目所有业务，为了提高开发效率可以直接引入ElementUI中的组件模板，如图18-8所示。

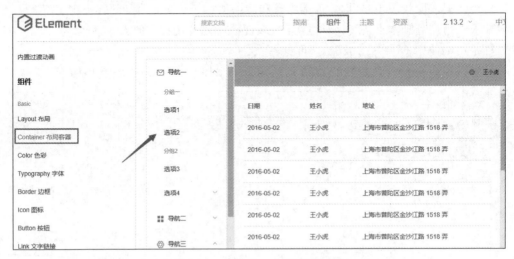

图18-8 Container布局容器模板

2. 创建首页

在 components 目录下新建 Index.vue 作为项目的首页,将图 18-8 中的菜单模板代码全部复制到 Index.vue 页面中。

菜单模板代码解释如下。

(1) el-container:构建页面框架。

(2) el-aside:构建左侧菜单。

(3) el-menu:左侧菜单内容,常用属性。

- :default-openeds="['1', '3']":默认展开的菜单。
- :default-active="'1-1'":默认选中的菜单。

(4) el-submenu:可展开的菜单,常用属性。

- index="1":菜单的下标,文本类型,不能是数值类型。

(5) template:对应 el-submenu 的菜单名。

(6) i 标签 class="el-icon-message":设置菜单图标。

接着修改路由配置,使用户能够直接访问首页,代码如下:

```
//部分代码省略
import Index from '../components/Index.vue'
export default new Router({
  routes: [
    {
      path: '/',
      name: 'Index',
      component: Index /* 首页 */
    }
  ]
})
```

这样就可以成功访问和图 18-8 一样效果的页面。本项目不需要这么多的菜单项,所以需要根据需求精简 Index 页面代码,代码如下:

```
<template>
  <div>
    <el-container style="height: 500px; border: 1px solid #eee">
      <el-aside width="200px" style="background-color: rgb(238, 241, 246)">
        <el-menu :default-openeds="['1', '3']">
          <el-submenu index="1">
            <template slot="title"><i class="el-icon-message"></i>导航一</template>
            <el-menu-item-group>
              <el-menu-item index="1-1">选项1</el-menu-item>
              <el-menu-item index="1-2">选项2</el-menu-item>
            </el-menu-item-group>
```

```html
            </el-submenu>
          </el-menu>
        </el-aside>
        <el-container>
          <el-main>
            <el-table :data="tableData">
              <el-table-column prop="date" label="日期" width="140">
              </el-table-column>
              <el-table-column prop="name" label="姓名" width="120">
              </el-table-column>
              <el-table-column prop="address" label="地址">
              </el-table-column>
            </el-table>
          </el-main>
        </el-container>
      </el-container>
    </div>
</template>
<!-- ... -->
```

首页简化后页面效果如图18-9所示。

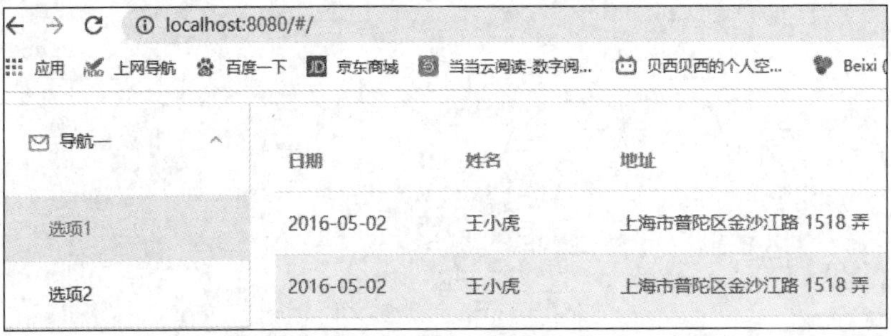

图18-9 首页

3．添加子页面

当单击"选项1"和"选项2"菜单，能够动态地加载子页面数据，所以在components目录下新建PageOne.vue（页面一）和PageTwo.vue（页面二），代码如下：

```html
<!-- PageOne.vue -->
<template>
<div>页面一</div>
</template>
<script>
    export default {
        name: "PageOne"
    }
```

```
</script>
<!-- PageTwo.vue -->
<template>
    <div>页面二</div>
</template>
<script>
    export default {
        name: "PageTwo"
    }
</script>
```

接下来在路由中配置首页的两个子页面，代码如下：

```
//部分代码省略
import PageOne from '../components/PageOne.vue'
import PageTwo from '../components/PageTwo.vue'
export default new Router({
  routes: [
    {
      path: '/',
      name: 'Index',
      component: Index, /*首页*/
      children:[/*子菜单*/
        {
          path: '/PageOne',
          name: '页面一',
          component: PageOne
        },{
          path: '/PageTwo',
          name: '页面二',
          component: PageTwo
        }
      ]
    }
  ]
})
```

配置完成后修改 Index 页面，使子页面内容能够被占位符接收，代码如下：

```
<!-- ... -->
<el-container>
    <el-main>
        <!--路由占位符用来接收子菜单内容-->
        <router-view></router-view>
    </el-main>
</el-container>
```

分别在浏览器中访问 http://localhost:8080/#/PageOne 和 http://localhost:8080/#/PageTwo 则会在页面中输出其对应内容，如图 18-10 所示。

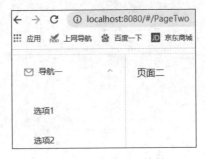

图 18-10　菜单页面效果

4．动态获取路由菜单

替换静态菜单为路由中的菜单项，并实现菜单和路由动态绑定，修改 Index 页面代码：

```html
<!-- 部分代码省略 -->
<el-aside width="200px" style="background-color: rgb(238, 241, 246)">
    <!-- 动态加载路由菜单,并实现菜单和路由动态绑定,那么必须在<el-menu>标签中添加router属性 -->
    <el-menu router :default-openeds="['0']">
        <!-- $router.options.routes 获取路由菜单 -->
        <el-submenu v-for="(item,index) in
            $router.options.routes" :index="index+''">
            <template slot="title"><i
                class="el-icon-message"></i>{{item.name}}</template>
            <!-- item2 就是二级子菜单,在 Vue 中跳转页面和 index 有关,item2.path 就可以实现菜单和路由动态绑定
                :class = "$route.path == item2.path?'is-active':''"在单击菜单时高亮显示
            -->
            <el-menu-item v-for="(item2,index2) in
                item.children" :index="item2.path"
                :class="$route.path == item2.path?'is-active':''">{{item2.name}}
            </el-menu-item>
        </el-submenu>
    </el-menu>
</el-aside>
```

接着我们希望用户首次访问时就显示页面一的内容，所以在路由中配置重定向，代码如下：

```
//部分代码省略
path: '/',
name: 'Index',
```

```
component: Index,
redirect:'/PageOne', /*首次访问时展示页面一的内容*/
```

至此动态加载路由菜单就成功了。

18.5 带分页数据查询

18.5.1 后端接口实现

修改 DeptController 类中的查询接口为带分页接口,代码如下:

```
@RestController
@RequestMapping("/dept")
public class DeptController {
    @Autowired
    private DeptRepository deptRepository;
    /* PageRequest 是 Spring Data JPA 封装的分页工具类。page 是当前页,size 是每页显示条数 */
    @GetMapping("/findAll/{page}/{size}")
    public Page<Dept> findAll(@PathVariable("page") Integer page,
        @PathVariable("size") Integer size){
        PageRequest request = PageRequest.of(page, size);
        return deptRepository.findAll(request);
    }
}
```

重启后端项目,在 Postman 中进行测试,结果如图 18-11 所示。

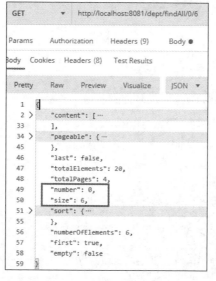

图 18-11 带分页数据信息

18.5.2 前端实现

1. 创建静态数据模板

首先使用 Element 组件 Table 表格构建静态表格数据模板,如图 18-12 所示,复制模板代码并覆盖 PageOne 页面,精简代码与数据库字段相对应。同时,引入分页模板代码如图 18-13 所示,当然大家可根据自己的喜好选择组件。

图 18-12　表格模板

图 18-13　分页模板

PageOne.vue 组件代码如下:

```
<template>
  <div>
    <el-table :data="tableData" border style="width: 100%">
      <el-table-column fixed prop="deptno" label="编号" width="150"></el-table-column>
      <el-table-column prop="dname" label="部门名称" width="120"></el-table-column>
      <el-table-column prop="loc" label="地址" width="120"></el-table-column>
      <el-table-column label="操作" width="100">
        <template slot-scope="scope">
          <el-button @click="edit(scope.row)" type="text" size="small">修改</el-button>
```

```html
        <el-button @click="delete1(scope.row)" type="text" size="small">删除</el-button>
      </template>
    </el-table-column>
  </el-table>
  <!-- 分页,总页数 = 总条数/每页显示的条数 -->
  <el-pagination
    background
    layout="prev, pager, next"
    :total="1000">
  </el-pagination>
  </div>
</template>
<script>
  export default {
    data() {
      return {
        tableData: [{
          deptno: '001',
          dname: '王小虎1',
          loc: '上海',
        }, {
          deptno: '002',
          dname: '王小虎2',
          loc: '上海',
        }, {
          deptno: '003',
          dname: '王小虎3',
          loc: '上海',
        }]
      }
    }
  }
</script>
```

2. 动态获取后台数据

在 PageOne 页面使用 Axios 请求后台数据接口，代码如下：

```html
<!-- 部分代码省略 -->
<!-- 分页,总页数 = 总条数/每页显示的条数 @current-change="page"单击当前页事件 currentPage 是当前页 -->
<el-pagination
    background
    layout="prev, pager, next"
    :page-size="pageSize"
```

```
          :total = "total"
          @current-change = "page">
</el-pagination>

<script>
  export default {
    methods:{
      page:function(currentPage) {
        this.$http.get("http://localhost:8081/dept/findAll/" + (currentPage - 1) + "/6").
        then(resp => {
          console.log(resp);  /* 在控制台查看分页信息 */
          this.tableData = resp.data.content;
          this.pageSize = resp.data.size;
          this.total = resp.data.totalElements;
        })
      }
    },
    data() {
      return {
        pageSize:'',
        total:'',
        tableData: [{
          deptno: '001',
          dname: '王小虎 1',
          loc: '上海',
        }, {
          deptno: '002',
          dname: '王小虎 2',
          loc: '上海',
        }, {
          deptno: '003',
          dname: '王小虎 3',
          loc: '上海',
        }]
      }
    },
    created:function(){/* 此方法和上面分页单击事件内容一样,它的作用是防止首次请求页面
无数据 */ this.$http.get("http://localhost:8081/dept/findAll/0/6").then(resp =>{
      console.log(resp);
      this.tableData = resp.data.content;
      this.pageSize = resp.data.size;
      this.total = resp.data.totalElements;
    })
  }
}
</script>
```

3. 测试

经过上面两步后,页面显示效果如图 18-14 所示。

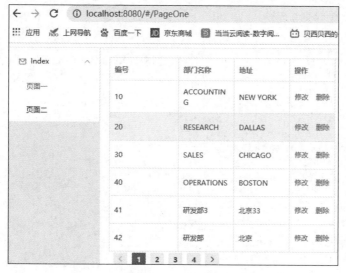

图 18-14 数据展示

18.6 部门员工信息的录入

23min

18.6.1 后端接口实现

在 DeptController 类中以插入方式添加接口,代码如下:

```
//部分代码省略
@PostMapping("/save")
public String save(@RequestBody Dept dept){
    Dept result = deptRepository.save(dept);
    if(result!= null){
        return "success";
    }else{
        return "error";
    }
}
```

18.6.2 前端实现

1. 修改菜单名称

在路由中修改菜单名称,并将组件 PageOne.vue 改名为 DeptManager.vue,将组件

PageTwo.vue 改名为 AddDept.vue，代码如下：

```
//部分代码省略
export default new Router({
  routes: [
    {
      path: '/',
      name: '部门管理',
      component: Index,          /*首页*/
      redirect:'/DeptManager',   /*首次访问时展示页面一的内容*/
      children:[                 /*子菜单*/
        {
          path: '/DeptManager',
          name: '部门查询',
          component: DeptManager
        },{
          path: '/AddDept',
          name: '添加部门',
          component: AddDept
        }
      ]
    }
  ]
})
```

2．修改 AddDept 页面

添加页面的布局，使用 Element 组件中的 Form 表单来实现，如图 18-15 所示。

图 18-15　Form 表单

页面布局成功后精简 AddDept 页面代码，并使用 Axios 将页面添加数据提交到后端项目添加接口，代码如下：

```html
<template>
  <el-form :model="ruleForm" :rules="rules" ref="ruleForm"
    label-width="100px" class="demo-ruleForm">
    <el-form-item label="部门名称" prop="dname">
      <el-input v-model="ruleForm.dname"></el-input>
    </el-form-item>
    <el-form-item label="地址" prop="loc">
      <el-input v-model="ruleForm.loc"></el-input>
    </el-form-item>
    <el-form-item>
      <el-button type="primary" @click="submitForm('ruleForm')">立即创建</el-button>
      <el-button @click="resetForm('ruleForm')">重置</el-button>
    </el-form-item>
  </el-form>
</template>
<script>
  export default {
    data:function() {
      return {
        ruleForm: {
          dname: '',
          loc: '',
        },
        rules: {
          dname: [
            { required: true, message: '请输入部门名称', trigger: 'blur' },
            { min: 3, max: 5, message: '长度在 3~5 个字符', trigger: 'blur' }
          ],
          loc: [
            { required: true, message: '请输入部门地址', trigger: 'blur' },
            { min: 3, max: 5, message: '长度在 3~5 个字符', trigger: 'blur' }
          ]
        }
      };
    },
    methods: {
      submitForm(formName) {
        this.$refs[formName].validate((valid) => {
          if (valid) {
            this.$http.post("http://localhost:8081/dept/save",this.ruleForm)
              .then(resp=>{
                if(resp.data == 'success'){
                  this.$alert('添加成功!', '消息', {
                    confirmButtonText: '确定',
```

```
                    callback: action => {/* 添加完成后跳转到 DeptManager 页面 */
                        this. $ router.push('/DeptManager');
                    }
                });
            }
        })
    } else {
        return false;
    }
});
},
resetForm(formName) {
    this. $ refs[formName].resetFields();
}
    }
  }
</script>
```

经过上面配置后,部门员工的录入功能就实现了。

18.7 部门数据编辑

18.7.1 后端接口实现

修改一般分两步:第一步先根据修改 ID 查询数据,第二步将数据修改后保存。所以在 DeptController 类中插入单条查询接口及修改接口,代码如下:

```
@GetMapping("/findById/{deptno}")
public Dept findById(@PathVariable("deptno") Integer deptno){
    return deptRepository.findById(deptno).get();
}
@PutMapping("/update")
public String update(@RequestBody Dept dept){
    Dept result = deptRepository.save(dept);
    if(result!= null){
        return "success";
    }else{
        return "error";
    }
}
```

大家可自行在 Postman 中测试。

18.7.2 前端实现

1. 建立修改页面

在 components 目录下新建 DeptUpdate.vue 组件,并在路由中配置,代码如下:

```
import DeptUpdate from '../components/DeptUpdate.vue'
//部分代码省略
children:[/*子菜单*/
  {
    path: '/DeptManager',
    name: '部门查询',
    component: DeptManager
  },{
    path: '/AddDept',
    name: '添加部门',
    component: AddDept
  },{
    path: '/update',
    component: DeptUpdate
  }
]
```

2. 绑定事件

在 DeptManager 页面修改绑定事件,并将行 ID 传递给 DeptUpdate 页面,代码如下:

```
//部分代码省略
edit(row) {
  this.$router.push({/*单击修改,跳转到修改页面并传递行 ID*/
    path:'/update',
    query:{
      deptno:row.deptno
    }
  });
}
```

3. 修改页面

修改页面可以借鉴添加页面的布局,在修改页面中首先通过 Vue.js 的生命周期 created() 方法在模板渲染之前获取后端数据,然后再渲染成视图。接着将修改的数据传递至后端进行保存,代码如下:

```
<template>
  <el-form :model="ruleForm" :rules="rules" ref="ruleForm"
    label-width="100px" class="demo-ruleForm">
```

```html
<el-form-item label="编号" prop="deptno">
    <el-input v-model="ruleForm.deptno" readonly></el-input>
</el-form-item>
<el-form-item label="部门名称" prop="dname">
    <el-input v-model="ruleForm.dname"></el-input>
</el-form-item>
<el-form-item label="地址" prop="loc">
    <el-input v-model="ruleForm.loc"></el-input>
</el-form-item>
<el-form-item>
    <el-button type="primary" @click="submitForm('ruleForm')">修改</el-button>
    <el-button @click="resetForm('ruleForm')">重置</el-button>
</el-form-item>
</el-form>
</template>
<script>
    export default {
        data:function() {
            return {
                ruleForm: {
                    deptno:'',
                    dname: '',
                    loc: '',
                },
                rules: {
                    dname: [
                        { required: true, message: '请输入部门名称', trigger: 'blur' },
                        { min: 3, max: 5, message: '长度在 3~5 个字符', trigger: 'blur' }
                    ],
                    loc: [
                        { required: true, message: '请输入部门地址', trigger: 'blur' },
                        { min: 3, max: 5, message: '长度在 3~5 个字符', trigger: 'blur' }
                    ]
                }
            };
        },
        methods: {
            submitForm(formName) {
                this.$refs[formName].validate((valid) => {
                    if (valid) {
                        this.$http.put("http://localhost:8081/dept/update",this.ruleForm).then(resp =>{
                            if(resp.data == 'success'){
                                this.$alert('修改成功!', '消息', {/*修改成功后跳转至 DeptManager 页面*/
                                    confirmButtonText: '确定',
                                    callback: action => {
```

```
              this.$router.push('/DeptManager');
            }
          });
        }
      })
    } else {
      return false;
    }
  });
},
resetForm(formName) {
  this.$refs[formName].resetFields();
}
},
created(){
  /* this.$route.query.deptno 接收添加页面传来的 ID,根据 ID 值调用后端 findById 接口
     查询数据 */
  this.$http.get("http://localhost:8081/dept/findById/" + this.$route.query.deptno).then
    (resp =>{
      this.ruleForm = resp.data;
    })
  }
}
</script>
```

经过以上几步配置后,部门数据的修改功能就实现了。

18.8 部门数据删除

18.8.1 后端接口实现

对于直接删除一般比较简单,只需在 DeptController 类中插入删除,代码如下:

```
@DeleteMapping("/delete/{deptno}")
public void delete(@PathVariable("deptno") Integer deptno){
    deptRepository.deleteById(deptno);
}
```

18.8.2 前端实现

在 DeptManager 页面,删除单击事件中使用 Axios 调用后端删除接口即可,代码如下:

```
delete1(row){
this.$http.delete("http://localhost:8081/dept/delete/" + row.deptno)
.then(resp =>{
    this.$alert('删除成功!', '消息', {
      confirmButtonText: '确定',
      callback: action => {
        window.location.reload();  /*删除完成后页面刷新*/
      }
    });
  })
}
```

至此，小型部门管理系统就完成了，相信大家对前后端分离会有直观的感受。后端负责实现 API 及业务逻辑，而前端可以独立完成与用户交互的整个过程，两者可以同时开工，不互相依赖，开发效率更高，而且分工比较均衡。

图书推荐

书 名	作 者
深度探索 Vue.js——原理剖析与实战应用	张云鹏
前端三剑客——HTML5＋CSS3＋JavaScript 从入门到实战	贾志杰
剑指大前端全栈工程师	贾志杰、史广、赵东彦
Flink 原理深入与编程实战——Scala＋Java(微课视频版)	辛立伟
Spark 原理深入与编程实战(微课视频版)	辛立伟、张帆、张会娟
PySpark 原理深入与编程实战(微课视频版)	辛立伟、辛雨桐
HarmonyOS 移动应用开发(ArkTS 版)	刘安战、余雨萍、陈争艳 等
HarmonyOS 应用开发实战(JavaScript 版)	徐礼文
HarmonyOS 原子化服务卡片原理与实战	李洋
鸿蒙操作系统开发入门经典	徐礼文
鸿蒙应用程序开发	董昱
鸿蒙操作系统应用开发实践	陈美汝、郑森文、武延军、吴敬征
HarmonyOS 移动应用开发	刘安战、余雨萍、李勇军 等
HarmonyOS App 开发从 0 到 1	张诏添、李凯杰
JavaScript 修炼之路	张云鹏、戚爱斌
JavaScript 基础语法详解	张旭乾
华为方舟编译器之美——基于开源代码的架构分析与实现	史宁宁
Android Runtime 源码解析	史宁宁
数字 IC 设计入门(微课视频版)	白栎旸
数字电路设计与验证快速入门——Verilog＋SystemVerilog	马骁
鲲鹏架构入门与实战	张磊
鲲鹏开发套件应用快速入门	张磊
华为 HCIA 路由与交换技术实战	江礼教
华为 HCIP 路由与交换技术实战	江礼教
openEuler 操作系统管理入门	陈争艳、刘安战、贾玉祥 等
5G 核心网原理与实践	易飞、何宇、刘子琦
恶意代码逆向分析基础详解	刘晓阳
深度探索 Go 语言——对象模型与 runtime 的原理、特性及应用	封幼林
深入理解 Go 语言	刘丹冰
量子人工智能	金贤敏、胡俊杰
Spring Boot 3.0 开发实战	李西明、陈立为
Flutter 组件精讲与实战	赵龙
Flutter 组件详解与实战	［加］王浩然(Bradley Wang)
Dart 语言实战——基于 Flutter 框架的程序开发(第 2 版)	亢少军
Dart 语言实战——基于 Angular 框架的 Web 开发	刘仕文
IntelliJ IDEA 软件开发与应用	乔国辉
Python 量化交易实战——使用 vn.py 构建交易系统	欧阳鹏程
Python 从入门到全栈开发	钱超
Python 全栈开发——基础入门	夏正东
Python 全栈开发——高阶编程	夏正东
Python 全栈开发——数据分析	夏正东
Python 编程与科学计算(微课视频版)	李志远、黄化人、姚明菊 等
Python 游戏编程项目开发实战	李志远
编程改变生活——用 Python 提升你的能力(基础篇·微课视频版)	邢世通
编程改变生活——用 Python 提升你的能力(进阶篇·微课视频版)	邢世通

续表

书　名	作　者
Python 数据分析实战——从 Excel 轻松入门 Pandas	曾贤志
Python 人工智能——原理、实践及应用	杨博雄 主编
Python 概率统计	李爽
Python 数据分析从 0 到 1	邓立文、俞心宇、牛瑶
从数据科学看懂数字化转型——数据如何改变世界	刘通
FFmpeg 入门详解——音视频原理及应用	梅会东
FFmpeg 入门详解——SDK 二次开发与直播美颜原理及应用	梅会东
FFmpeg 入门详解——流媒体直播原理及应用	梅会东
FFmpeg 入门详解——命令行与音视频特效原理及应用	梅会东
FFmpeg 入门详解——音视频流媒体播放器原理及应用	梅会东
Python Web 数据分析可视化——基于 Django 框架的开发实战	韩伟、赵盼
Python 玩转数学问题——轻松学习 NumPy、SciPy 和 Matplotlib	张骞
Pandas 通关实战	黄福星
深入浅出 Power Query M 语言	黄福星
深入浅出 DAX——Excel Power Pivot 和 Power BI 高效数据分析	黄福星
从 Excel 到 Python 数据分析：Pandas、xlwings、openpyxl、Matplotlib 的交互与应用	黄福星
云原生开发实践	高尚衡
云计算管理配置与实战	杨昌家
虚拟化 KVM 极速入门	陈涛
虚拟化 KVM 进阶实践	陈涛
边缘计算	方娟、陆帅冰
LiteOS 轻量级物联网操作系统实战（微课视频版）	魏杰
物联网——嵌入式开发实战	连志安
HarmonyOS 从入门到精通 40 例	戈帅
OpenHarmony 轻量系统从入门到精通 50 例	戈帅
动手学推荐系统——基于 PyTorch 的算法实现（微课视频版）	於方仁
人工智能算法——原理、技巧及应用	韩龙、张娜、汝洪芳
跟我一起学机器学习	王成、黄晓辉
深度强化学习理论与实践	龙强、章胜
自然语言处理——原理、方法与应用	王志立、雷鹏斌、吴宇凡
TensorFlow 计算机视觉原理与实战	欧阳鹏程、任浩然
计算机视觉——基于 OpenCV 与 TensorFlow 的深度学习方法	余海林、翟中华
深度学习——理论、方法与 PyTorch 实践	翟中华、孟翔宇
HuggingFace 自然语言处理详解——基于 BERT 中文模型的任务实战	李福林
Java+OpenCV 高效入门	姚利民
AR Foundation 增强现实开发实战（ARKit 版）	汪祥春
AR Foundation 增强现实开发实战（ARCore 版）	汪祥春
ARKit 原生开发入门精粹——RealityKit + Swift + SwiftUI	汪祥春
HoloLens 2 开发入门精要——基于 Unity 和 MRTK	汪祥春
巧学易用单片机——从零基础入门到项目实战	王良升
Altium Designer 20 PCB 设计实战（视频微课版）	白军杰
Cadence 高速 PCB 设计——基于手机高阶板的案例分析与实现	李卫国、张彬、林超文
Octave 程序设计	于红博
Octave GUI 开发实战	于红博
全栈 UI 自动化测试实战	胡胜强、单镜石、李睿